The Omega Matrix

The Omega Matrix

Putting Ecological Rudder to Global Enterprise

Ralph H. Miner

Writers Club Press

San Jose New York Lincoln Shanghai

The Omega Matrix
Putting Ecological Rudder to Global Enterprise

Writers Club Press
an imprint of iUniverse.com, Inc.

For information address:
iUniverse.com, Inc.
5220 S 16th, Ste. 200
Lincoln, NE 68512
www.iuniverse.com

ISBN: 0-595-14386-5

Printed in the United States of America

Dedicated
To My Wife

Rosemary

CONTENTS

PROLOGUE

The Drift From Coherence

We have now reached that pivotal point in history where we must vigorously challenge one of two radically divergent perspectives of the world's future.

The widest and most alarming perspective is held by the massive community of technologists and industrialists. Their view holds that the future world economy will be one of projecting an already buoyant change-driven economy to still greater glory. They tend to hold this view while wearing blinders to existing inhumane conditions in America and throughout the world.

The more realistic view, on the other hand, is held by an overwhelmingly outvoted minority—the environmentalists and the broadly backgrounded researchers of the future. Their view is that during the next half-century, with another four-billion people added to the world's present population of six-billion people, we are headed for an ecological disaster and an inhumane quality of life far worse than an already critically diminished state, accompanied by a galloping increase in world hunger.

The OMEGA matrix—Omni Matrix, Economic Growth Analysis— seeks to challenge the prevailing view and to present a forum within which these two divergent forces can, in time, be reconciled to a coherent world economy in a humanely balanced ecosystem.

At the outset, however, it must be noted that the OMEGA matrix can not be defined in a brief paragraph or two. We shall nonetheless define the structure of the OMEGA acronym. This book is not a one-dimensional work addressed to a single area of concern. It is multi-dimensional and presumes to measure up to the definition of the term "omni", a combing form meaning all; totally; everything. Then the term "matrix"—that in which anything originates, develops, takes shape or is contained; an array of numbers or terms to facilitate the study of relationships. Hence, Omni Matrix everything shaped in terms to facilitate the study of relationships and within which something else originates or takes shape.

Then the term "economic growth analysis" not as economic theory but economic growth as it is generally perceived—growth of population, growth of resource demands and growth of services.

Hence, Omni Matrix, Economic Growth Analysis…OMEGA.

In order to unfold OMEGA in all of its basic aspects, we must first address five key issues to be reckoned with before we can construct an OMEGA aimed toward global enterprise in the context of a survival oriented global ecosystem.

But before prefacing those five key issues it is necessary to penetrate a heavy mist now obscuring a clear view of the larger problem. As the course of human history virtually explodes into a hypercomplex new age, this new age is accompanied by an obscuring mist in the form of a torrential flow of random, grossly unstructured information. This has engulfed informed citizens and professionals in all fields in dangerously fragmented thinking, an obscuring mist leading us ever deeper into our present economic and ecological dilemma. Moreover, this is further aggravated by the volume of random information and knowledge doubling every five years. One way or another all of the players in the global enterprise are involved in reckoning with this Niagra of critically unstructured information. Relevance often lost in a sea of nonrelevance.

Now what sort of schism in our general thinking would foster our present inundation in such massive amounts of low quality and often useless information?…Or, indeed, what kind of mind-set perpetuates suffocation of many of the contemporary writings on new approaches to our incoherent ecosystem?

The answers to these questions are of course enormously complex and largely what this book presumes to address. While not easily dealt with, the basic answer to these questions lies in an absence of a manageable macro-system structure of the world economy. Such a macro-management model would serve as the foundation for the management of all economic growth and sociological factors affecting the well being of mankind.

Included in this view is not only a desperate need for a clearly understandable structure of the known dimensions of the world economy, but a structure which includes the dimensions of the rapidly deteriorating global ecosystem and, in turn, a macro-management structure, hopefully OMEGA, capable of responding to those long neglected pivotal issues which have caused the present untenable world crisis. Those five key issues are:

- Religion versus science yet unreconciled.
- Natural and man-made environments clashing.
- Entropy law ignored—ecosystem paradigm nonexistent.
- Economic theory unable to reckon with today's needs.
- Time compression largely ignored—pile up of events.

It is appropriate to prologue these key issues since they are brought to light and enlarged upon throughout the text of the book.

In our approach to religion versus science we respond to constitutional principle and ethical parameters. This enters poignantly into each of the key issues. Drawing on the philosophy exercised by the framers of the Constitution of The United States, this book, seeking to urge wide application of constitutional principle and ethical parameters will embrace God, the noumenon of Creation as the founders platform for the constitution, but will place religion in a pragmatic context. (The theistic aspects of the constitution will be briefly enlarged upon in Chapter One.)

Indeed religion and science must join one another in a new world paradigm. The distinguished Chairman of the Executive Council of the prestigious California Institute of Technology, Dr. Robert Millikan, offered at mid-20th century a profound statement on the relationship of religion and science. Now, more than half-a-century later, it is clearly evident that we have drifted away from his view. Nonetheless his statement stands today as a strong admonition and timeless guide post. Millikan said:

"Human well being and all human progress rest at the bottom upon two-pillars, the collapse of either one of which will bring down the whole structure. These two pillars are the cultivation and dissemination throughout mankind of (1) the spirit of religion, (2) the spirit of science."

Of course. Dr. Millikan said a great deal more in his amplification of that basic statement. But for prologue purposes, the essence of his platform will suffice. He also tied in and amplified a quote of Philosopher Alfred Whitehead. In his brief but poignant summary of religion, Whitehead said:

"Religion is world loyalty."

That statement, reflective of God's word, is a pragmatic credo which underlies the approach to this book. Whitehead's term also tends to neutralize the myriad facets of world religions, often at cross purposes with one another.

Correlative to this is a quote of Dr. Millikan's enlarged reference to a joint statement on "The Relations of Science and Religion" by a distinguished group of scientists, religious leaders and men of affairs made over three-quarters-of-a-century ago. It is noteworthy that even so early in the twentieth century their statement was prompted, they said, by the "observation of on-going controversies in which science and religion were presented as irreconcilable and antagonistic domains of thought". Hence, as they were gravely concerned over the wrongs of this rather widespread view, here is their penetrating joint statement:

"The purpose of science is to develop without prejudice or preconception of any kind, a knowledge of the facts, the laws and the processes of nature. The even more important task of religion, on the other hand, is to develop the consciousness, the ideals and the aspirations of mankind. Each of these two activities represents a deep and vital function of the soul of man, and both are necessary for the life, the progress and the happiness of mankind."

"It is a sublime concept of God which is furnished by science, and one wholly consonant with the highest ideals of religion, when it represents Him as revealing Himself through countless ages in the development of the earth as an abode for man and in the age-long inbreathing of life into its constituent matter, culminating in man with his spiritual nature and all his Godlike powers."

But what is happening today at this dawn of the third millennium?...Both pillars—"the spirit of religion" and "the spirit of science" in the context of "world loyalty" are clearly revealing collapse-threatening major cracks and deep fissures. The sweeping deleterious effects of this are most poignantly summed in one all embracing world wide situation—the incipient deterioration of the global ecosystem...and all that is implied by that in terms of human suffering and lack of spiritual equity.

The explosive scientific, technological and economic growth of the twentieth century wildly exceeded the expectations of even the most adventuresome techno-economic growth forecasters of mid-century. But no properly authorized constituency existed with a charter to build a communication bridge between the pillars of religion and science in the context of world loyalty. Had such a communication bridge existed, it quite likely would have been seen early in the second half of the twentieth century that the two environments we know today—natural and man-made-could no longer be endured. That is, no longer endured in the light of the wide gulf in motivation between the so-called protectors of the natural environment and the industrial advocates of extensive man-made environments.

At this juncture in history there should be only **one environment**—a harmonious integrated ecosystem embracing both natural and man-made environments. That one environment would be known as the man-engendered global ecosystem. Hence, under a properly formulated macro-management system which would integrate presently antagonistic industrial and environmental factions, a new order of things could be realized.

From this, in a climate of "world loyalty" and all that is implied in that, we could bring never-to-end exponential economic growth and a diminishing natural environment toward ultimate realization of **one** orderly global environment. In turn, harmonious accommodation of all the peoples of the world would eventually emerge in a climate of physical well being and spiritual equity...Idealistic? Perhaps. But a new synthesized way of thinking about the enormity of the world ecosystem problem must be launched soon in order to make meaningful inroads into the process of integrating all of the agencies involved, not the least of which would be those agencies involved in the geopolitical arena. Even in that arena, while there is considerable international effort to bring ecosystem issues forward, there is still lacking an integrated operational paradigm within which to focus goals and their implementation.

Not to preempt the OMEGA matrix text, but it must be underscored that focusing on the world ecosystem as one man-engendered environment by no means infers a limited approach to global enterprise. Moreover, focusing on the world ecosystem via the new paradigm—again, in the context of world loyalty—brings every conceivable scientific, ethical and economic growth act into cause and effect analysis with not only relevant resolution but many opportunities for expanded enterprise.

Now, all of this is not to say that the environmental people and the science and technology people are not presently making important contributions to a better environment. Some good things are happening. But thus far their efforts are too piecemeal and grossly unstructured in the context of the whole of things. In turn, we fall further behind every year in properly addressing the whole of the problem.

Of the many elements involved in addressing a new order of things we need perspective on the underlying cause of the open warfare between the ecological purists and the industrial advocates of maximized economic growth. This tug-o-war between preservation of the natural environment and the industrial growth associated with containment of an ever-burgeoning world population must not continue. As noted, some good accomplished, yes. But bringing these two adversarial factions together in indisputable quantitative and qualitative terms requires, in addition to strong ethical input, a recognition of the oldest and currently most ignored law of modern science. And that law is the law of entropy.

The entropy law is the second law of thermodynamics and states that matter and energy can be changed in only one direction, that is, from usable to unusable. In the process of usable to unusable there is always some waste. Oversimply for now, that waste is known as entropy. Our primary goal must be to reduce unacceptably high rates of entropy on a global scale. Analytical perspective on this is found in what may be defined as the "energy axis"—energy converted plus entropy rate equals ecosystem status. It will also be shown that the energy axis is the key girder in building a bridge between the pillars of religion and science.

Einstein said the entropy law is "the premier law of all science"...Sir Arthur Eddington referred to the entropy law as "the supreme metaphysical law of the universe"...Jeremy Rifkin, distinguished social commentator who has written profoundly on the myriad scientific and social dimensions of entropy, states that: "The entropy law will preside as the ruling paradigm over the next period of history." (Rifkin's book, "Entropy", is a must read!)

While one can quickly agree with Rifkin on this, it is not happening to any marked degree. Yet his profound and compelling advocacy was written two decades ago.

Hence, to move toward bringing the entropy law to a presiding role, we must first overcome the misguided environmental precedents fostered by the ecologists a few decades ago. In their headlong pursuit of environmental deficiencies, they virtually ignored the causal significance of the entropy law. In that near past, the general public did not even know of the term "ecology", let alone the term "entropy". Yet, through myriad media-supported protest marches and overly exploited oil stained sea gulls, radical environmentalists soon turned ecology into a household word. And in that misguided environmental campaign, the public was led astray of exposure to the larger problem.

While entropy is daily analyzed by scientists and engineers in solving relatively low order problems, the vital and overriding entropy law has no presiding posture whatsoever in the public domain...nor, even worse, in the highly influential science and technology community.

Now, all this is not to put down the substantial accomplishments of the environmental movement. Their contributions to at least some pollution reduction and their bringing about more energy efficient machines, buildings and appliances is most worthwhile. Yet tunnel vision dominates the environmental perspective. With vigor the environmentalists, among other things, seek to arrest new infrastructure development and even attempt, often succeeding, to stifle land use zoning changes required to satisfy legitimate economic growth needs.

On the other hand, the environmentalists are sometimes correct in attempting to stop certain development when such development threatens to further choke existing overstressed infrastructure. But the environmentalists are inconsistent and often whimsical, lacking wide-angle ecological vision or, even worse, there are too many unrestrained environmental charlatans who bend environmental impact reports to serve political ends, thus facilitating unethical development.

The above example applies to the entire spectrum of ecological issues facing society today. Whether seeking pollution abatement, preservation of the natural habitat of the dusky footed wood rat, the spotted owl or the banana slug, fragmented approaches to ecological concerns are not the way to proceed.

And this is not intended to single out the relatively small minority of environmentalists. It will be made abundantly clear that an overwhelming majority of industrialists share to a far greater extent the lack of wide-angle ecological vision and its deleterious consequences. But the industrialists are often right, too, when fragmented environmental perspectives are found unreasonable in the light of realistic containment of the needs of people.

Wide-angle ecological vision, on the other hand, could resolve this dilemma by weaving a fabric of consistency over the ecosystem as a whole. Such would embrace with ecosystem balance the combined needs of inexorable economic growth and reasonable protection of natural wonders.

Hence, the basic approach to the lack of wide-angle ecological vision and the degree of environmental charlatanism existing today is to see the entropy law placed in a yet nonexistent high order controlling paradigm over which it can indeed preside.

Briefly outlining such a paradigm: In hierarchical order, ecosystem problems would be fed through (1) Constitutional **Principle** reinforced by appropriate ethical parameters input. Then flow through (2) A world-oriented statement of humane **Purpose.** And then flow into (3) A corresponding macro-management **Precept** which responds to the entropy law in the context of the economic growth demands placed on the **Geosphere**, the **Biosphere** and the **Sociosphere.** (Full enlargement of this will be offered in the OMEGA matrix text in Part III: Putting Ecological Rudder to Global Enterprise.)

To underscore a key aspect of the proposed new paradigm, a timeless, forever unchanging basic structure is being offered. In such an all-embracing entropy premised structure, numbers would come and go but, once mandated, the basic **analytical structure and flow pattern would remain forever unchanged** and, as will be offered, an

OMEGA matrix, or something akin to it, forever sifting the relevant from the nonrelevant.

Entropy is an all-pervasive law, with inferences reaching far beyond the energy efficiency improvements of household appliances and pollution reduction. Moreover, it is a law bearing not only on the vast mechanical system overloads confronting us, but on emotional and sociological overloads as well.

Consider Los Angeles and its urban sprawl, for example. That city is an entropic disaster and growing worse. Yes, it works…sort of. But the price paid in inordinately high entropy rates through patchwork expansion and redevelopment—putting blow out patch upon blow out patch, so to speak—has long since passed the entropic saturation point. Saturation obtains not only as an inefficient conglomeration of urban and sub-urban systems, but in its physical and emotional saturation of millions upon millions of people as well…And New York is probably still worse as it faces the entropic nightmare of a 92-Billion dollar refurbishing bill on its century-old infrastructure.

The entropy law approach? The above urban situation and others not unlike it throughout the world, and many still worse, must address the "entropic relief valve". This means the implementation of great new metropolitan centers from scratch, strategically placed in appropriate geographical and inter-urban logistics positions. We will show, in the text of the book, the significance of this in achieving massive life-enhancing entropy reductions while at the same time properly containing never-to-end exponential economic growth. This discussion will include, of course, many logical entropy reduction opportunities in the sub-systems of economic growth—utilities, commodity production, transportation and so forth. Now, this idea in itself has been suggested by others. But to the best of the writer's knowledge, a sweeping entropy law paradigm such as the OMEGA matrix—embracing the ethical parameters of constitutional principle, a world oriented humane purpose and a macro-management precept for integrating the whole of things, together with an action structured ethical feedback system—has not been offered.

All this idealistic? Yes. Yet, when the basic concept gains a foothold in the research community at large, idealism will be tempered by further refinement of the concept. In any case, there is every possibility that such a concept will catch hold as that education program which ought to be imbued in the young minds and psyches of those people who will one day manage the transnational new world we hope to perpetuate. Actual implementation of such a program in the present scheme of things would, of course, occur on an evolutionary basis, maturing over a period of time.

At this point in the prefacing remarks, it is necessary to return briefly to the antagonistic industrialists and environmentalists. There is little doubt that in many cases the industrialists need to clean up their act, reducing pollution levels and so forth. The environmentalists, on the other hand, need to know a great deal more about what is involved in legitimate economic growth before precipitously rushing in to obstruct that growth.

Industrialists must respond to an entropy premised macro-management structure and environmentalists need to be educated to respond to that same structure. While both of these adversarial factions are guilty of sins of omission and commission, they must be brought together—and this bears repeating—on the moral and ethical premise of **one** entropically controlled, man-engendered global ecosystem.

Moving toward the close of our prologue, a few caveats are offered as they bear on stimulating response to the urgency of going ahead with an all embracing system based on the entropy law.

We are now at a point in history where the fierceness of profit motivation is not only suffocating much of the effort to correct the ecosystem, but fosters an ignoring of the portent of the explosive economic growth dynamics of the past few decades. The great exponential curve of population growth in flowing time and the hyper-exponential curves of corresponding subsistence dynamics, now thrusting near vertically, have not sufficiently touched the minds of most professionals nor the community of predatory capitalist groups, at least insofar as their response to the world ecosystem and the transnational global economy is concerned.

As a result, the generation of the glut of nonrelevant information we now endure continues on the rise. Moreover, too many studies are static studies and are not dynamically tied in with the diverse operational systems comprising the present day wobbly whole of things.

Too many professionals generally fail to recognize that the economy and the ecosystem as a whole is comprised of human beings on many levels together with myriad operational systems serving them. Many of these systems, of course, are global in their transnational structure. Humans interact with these systems and with each other, and all of this in the unceasing and indivisible dynamic flow of time. It is in this relentless dynamic flow of time that qualitative values arise which often tend to defy inclusion in the quantitative investigation. Many analysts tend to shun these hard-to-integrate qualitative values and compile their data on static quantitative bases. And it is in these terms that while the arithmetic may be accurate, it is accurate only within narrow, incompletely defined boundaries...and therefore not meaningfully useful.

For example, consider the economist's usual reporting mode on energy output—the backbone of all economic activity. Now, economics may well be defined by economists as "the discipline dealing with the production, distribution and consumption of wealth", but look at the manner in which total energy output is reported. Energy is measured and reported in kilowatts, without regard to, or insufficient regard to, origins of energy conversion. And certainly not with discernible concern for the constant dynamic shifts in energy production resource modes...not to mention the accompanying extreme variations in entropy rates and ecosystem impact. Yes, some research has examined this qualitative situation, but without recourse to the kind of entropy-premised total system paradigm that would bring focus to the overall situation.

Indeed, consider the hodge-podge of energy conversion media now permeating the world economy. We have, on one hand, deeply entrenched in our economy, the fossil fuel based energy media—oil, coal, gas, wood—with their enormously high entropy rates and attending negative ecological effects. On the other hand, we now have emerging dynamic based, very low entropy systems such as solar, wind, tidal and geothermal media. These, together with with the nuclear energy situation will be discussed in subsequent chapters. The fierce combination of these three energy media groups—fossil fuel based, dynamic based and nuclear based–is fostering a growing degree of unmonitored indiscriminate economic growth. The traditional reporting modes of the economics community not only do not help make a case for the economic portent of all this, but they do not provide data readouts which facilitate analysis of ecosystem inferences on an interdisciplinary basis. Moreover, due regard is not given the fact that various operational systems flow in the mainstream of time at different rates and interact with each other in ways difficult to measure. (Drawing on the timeless philosophy of Henri Bergson regarding time flow and duration, we will explore this aspect in some detail in Part II.) In short, economic growth analyses are performed almost without regard for the hard-to-quantify qualitative factors involved in time-flow-sensitive operational interplay—interplay among diverse national and transnational systems of machines and technologies and the people they serve.

In any case, it is inevitable that whatever entropy law paradigm might be developed—the OMEGA matrix, or something akin to it—will have great difficulty becoming an integral part of economic theory and policy...if, indeed, it is even possible. The transnational economy and its ecology is here to stay, along with all that is implied in melding this to one world ecosystem.

The explosive world population growth, accompanied by still more explosive growth of utilities, transportation and commodities systems—together with heavy technological innovation of these systems at uneven rates—raises serious questions as to the capacity of economic theory, old or new, to contend with this situation. Obviously, this calls for radical changes in favor of new methodology for economic growth analysis—a **clearly understandable** methodology.

To be sure, a grand synthesis of the whole of things is required in order to arrive at any kind of functioning economic growth analysis system, let alone a system for controlling piecemeal applications of economic theory. Moreover, present economic theory becomes increasingly obscure to those government leaders and private industry executives who make the decisions which steer world development. We must develop a clearly structured, interdisciplinary economic growth analysis system which will optimize the decisions of those leaders.

In turn, it is more than likely that classical economic growth analysis would be confined to the solution of various narrowly circumscribed problems. This rather than permitting some unproven new economic theory which only purports to be an all embracing coherent system. Indeed, referring to informed current commentary on the subject, one does not get the feeling that the classical economic theorists have a clue as to the development of a new economic theory capable of embracing the transnational economy and its explosively changing world ecosystem. (For compelling and extensive enlargement on the futility of classical economic theory in today's world, see Peter Drucker's "The New Realities".)

Enter, then, The OMEGA Matrix—

Omni Matrix, Economic Growth Analysis.

First, in the light of the discussion thus far, a few prefacing remarks are offered on what OMEGA is and what it is not.

As noted earlier, OMEGA does not purport to be economic growth analysis in any classical sense. Semantically, the term "economic growth" refers only to its widely accepted meaning in its most general sense, namely growth—growth of population, support systems and services—which is generally understood to mean economic growth.

The OMEGA Precept will be delineated as a plan for ethical, technological and operational conduct in which a viable world ecosystem eventually could be attained. The actual construction of the three-dimensional OMEGA matrix will be described, which is an analogue of the prime components of the world economy and its ecosystem. As such it can be found a panacea for information collecting, validating and processing on a macro-scale never before attempted. Attempted, that is, in a manner which can not only assure a high degree of information relevance, but which is **comprehendible** to informed citizens and all professional disciplines alike.

Underscoring the vital importance of a timeless and unchanging basic analytical structure for sifting out the nonrelevant and processing all relevant information, the prefacing remarks close with a summary of the primary needs-of-mankind issues facing us in the new century.

First, world population has not yet peaked at its present level of nearly six-billion people. Indeed, projections by the United Nations indicate an increase to ten billion or more people during the next half century. Yet, in the **last** half century, the key world food staple, grain production, increased by only 19% in the face of a 132% increase in world population. The portent of this for the next half-century is obvious.

Then, of course, there is the water consumption demand required to fulfill this continuing explosion of population and attending subsistence demands. Spreading water scarcity is probably the most underrated resource issue in the world today, notwithstanding its growing obviousness. Rivers are being drained and water tables are falling on every continent. Great rivers such as the Nile, the Yellow and our own Colorado now have little water left when they reach the sea. The Water Management Institute projects that in only a quarter-century, a billion people will be facing absolute water scarcity—equivalent to nearly four times the U.S. population.

And, of course, there are myriad related concerns. Energy demands, together with raw materials, systems and commodities demands growing at an exponentially greater rate than population growth…Then the no longer tolerable entropy-premised global warming trend…And, far from least, underlying all of the above, the horrendous problem of waste disposal. Waste management research finds that world wide garbage, sewage and industrial waste disposal is presently far from under control, let alone the spectre of the overall waste disposal situation by mid-century.

And, of course, there is the enormous burden of education at all levels. Not only for the projected increase of 93% in child age population over the next fifty years, but the mandatory requirement for continuing adult education at all levels in the face of the ever-increasing complexity of new technology.

All of the above overwhelming?…It need not be. The pivotal answer, though not so simply, lies in sifting relevant information and executive actions from the present crushing burden of incoherent nonrelevancies. All future economic growth analyses should proceed with this in mind.

As a closing comment to the prologue, the reader will find this book a strong advocacy for the macro-world of the "Clock Culture"—of real time as we daily live it. It is **not** an advocacy for the micro-world of "Computopia"—the nanosecond life of unrestrained information flow in which response to relevant human need is only happenstance, and in which change for change's sake becomes the new ethic.

The keynote of this book is the seeking of relevance to the conditions herein prologued, drawing upon the **relevant** wonders of micro-science without being ruled by a **nonrelevant** micro-ethic which allows micro-creative processes to roam freely and find their own order.

Indeed, let us arrest the drift from coherence and center on analysis of relevance to the human condition—coherence through a relevance-ordered structure for analyzing the world in the context of how it ought to work in never-to-end economic growth—putting ecological rudder to global enterprise.

*   *   *

Part I
PRINCIPLE

CONSTITUTIONAL PRINCIPLE...
NOT TECHNO-ANARCHY

"When right I shall often be thought wrong by those whose positions will not command a view of the whole ground. I ask your support against others who may condemn what they would not if seen in all its parts..."

Thomas Jefferson

CHAPTER ONE

Constitutional Origins Vs. Ecosystem Chaos

The three chapters of Part I offer a dissertation on principle, together with moral and ethical parameters, as an overview—a distillation of essences, a sweep in time without ponderous historical dialogue. Offered is the essence of where OMEGA is coming from.

The ultimate basis on which the Constitution of The United States rests, and the accepted philosophy in American law, is the assumption of the existence of a Supreme First Cause—God, the Creator of us all.

Fundamental and useful is the notion of a necessary Being—God, who originates all else. Yet, this underlying constitutionl principle, which contains rational proof of the existence of God and has had profound influence on civilization and the evolution of the great body of law we know today, is threatened by the contradiction of antithetical thought and action. The insidiousness of this is now seen in a growing dominance of antithetical thought, not only by those engaged in antithetical practices, but by a still larger number of those who profess not to be in opposition to constitutional principle.

How then did this alarming situation come to pass?

There seems to exist today a surprising preponderance among professionals in many disciplines toward a tendency to recognize constitutional law on one hand, while on the other hand offering resistance to the theistic origins of the constitution. As new legislation is developed today in order to reckon with modern times, the theistic aspects exercised by the founders over two-hundred years ago seems to be set aside as not relevant to the implementation of new law.

Many years ago, the writer was exposed to a superb and lengthy lecture on "Jurisprudence and Legal Institutions" by Professor William F. Roemer of the University of Notre Dame. A few passages are quoted from a published copy of that lecture. These passages make abundantly clear that the Constitution of The United States and the theistic philosophy underlying it can not be separated in time merely because the constitution has been established. Continuing response to its theistic foundation is more vital today than ever.

Under one of the sub-headings of his lecture, "Basic Truths of Jurisprudence", Dr. Roemer said:

"In American law, the accepted philosophy upon which the Constitution of the United States rests as its ultimate bedrock is theistic, assuming the existence of a supreme first cause or Deity. This truth is the first postulate in the study of jurisprudence."

"This postulate of God's existence, as a personal and intelligent Creator of the universe, who is the Supreme Lawgiver by title of ownership over men and their property, was incorporated as evident truth into the foundation of our Federal Constitution. Earlier it was briefly in the second paragraph of the Declaration of Independence which enunciated the right of Americans to form a new political state."

He went on to say in another sub heading of his lecture, "Objective Ethical Basis of Law":

"No man or men can disregard the Maker of men and God's law in the formation of regulations for his fellowmen, lest they step outside of the bounds set by the higher Natural Law."

"Using the language of philosophy, we find proof of the Natural Law in several sources, among them metaphysical premises. The metaphysical argument utilized by Philosophical Jurisprudence is based squarely on the ground that the all-wise Creator could not be conceived as indifferent as to whether His creatures did or did not attain the end for which He designed for them in His Eternal Plan. For such an indifference would imply a lack of intelligence and capricious folly in God, which is a contradiction of Divinity."

"Once having constituted man a free-agent by endowing him with the faculty of choice between rationally known alternatives of conduct, God would direct and manage man only by moral law, which binds his nature with a moral obligation."

Drawing on the inspiration of Dr. Roemer's lecture, the causal essence of the present antithetical outlook is this: At the time of the drafting of the Federal Constitution, with proper homage to "In God We Trust" and "E Pluribus Unum" being proffered by our leaders, the industrial revolution was concurrently moving into place. Too, at that time, modern science was very much in its infancy while at the same time finding myriad new demands placed upon it.

Before the upward thrust of a burgeoning new technology evidenced itself, together with its promise of explosive economic growth, we found the political statesmen, the legislators, the physical scientists, the economists and a whole new breed of industrial technologists working essentially in harmony with one another, give or take relatively modest disputes. At that time there existed a mutual endorsement of investigative methodology and, for the most part, philosophical concurrence on the approach to developing a great new nation.

In that early period of The Great Concurrency—the new Federal Constitution, the Industrial Revolution and an emerging Modern Science—" E Pluribus Unum" was not an abstraction and "In God We Trust" not merely a motto stamped on our coinage.

All too quickly, however, the ensuing decades saw the promise of a burgeoning rate of economic growth and the dominance of profit motive take over the industrialists' general line of thinking. Soon, a bifurcation of thought set in. The constitutionally oriented investigators of morality and ethical parameters moved in one direction with an essential purity in their methodology. On the other hand, the promulgators of industrial technology, increasingly dominated by the profit motive and the attending requirement for accelerated economic growth, narrowed their investigative methodology to a point where constitutional principle and the search for ethical parameters drifted from immediate view, yielding a massive antithetical effect.

Thus, the practitioners of constitutional principle, insofar as their impact on the operational mainstream of of economic growth is concerned, have become one of the more suppressed minorities in history. While doing much investigative work on the frontiers of socioeconomics and the human condition, they are not making their message felt to a point of regularly implementing adequate corrective actions. Those professionals who have moved us down a techno-economic path of antithetical effects outnumber the active advocates of constitutional principle many thousands to one. The advocates of constitutional principle and ethical parameters are simply overwhelmingly outvoted…though, fortunately, the fundamental core of constitutional principle perseveres.

Hence, we shall emphasize a basic philosophical approach, predating the Federal Constitution by many centuries but brought into practical modern focus by the framers of the constitution. This is an approach which, with little amplification of basic structure, applies today to all professions. This is the investigative method of philosophical jurisprudence—the practical science which investigates the nature, origin and function of law. It is grounded on the true metaphysical principle of God as the Creator of us all, inferring that He is the origin of our guidance in the postulation of ethical parameters for the conduct of life.

We have elected to ascribe optimum investigative method to philosophical jurisprudence not only because of living proof of its viability as found in the constitution, but because law is a generic term embracing physical science as well as social phenomena. Parenthetically, it is urged that science and engineering curricula come to require courses in this subject.

This philosophy, predating modern science, can rightly be acknowledged as providing the very foundation stones of modern science and the scientific method, yet reaching far beyond the scientific method as it is generally practiced today. With some exceptions, physical scientists, engineers and technology at large stop far short of the investigative methods of philosophical jurisprudence.

As to the basics of this philosophy, and of the scientific method, much will be said later of inductive and deductive analytical processes, the synthesis of wholes and hierarchical orders of things.

At this point, in reaching for the truth and the whole of things, we shall simply highlight a critical gap between those professionals oriented to constitutional principle and those professionals who either overtly oppose or whose actions manifest opposition to constitutional principle and ethical parameters.

The practitioners of philosophical jurisprudence are sometimes contestable, but it is method we are subscribing to here. It is a method which opens wide avenues of awareness to higher orders of things…and **equal access** to decision making by all participants involved in a given situation.

In developing the ethical and moral parameters of a situation, whether a professional in the field of religion, physical science, technology, legislation or whatever, the real investigator seeks to **know truly the outermost boundaries** of that situation and its peripheral inferences. Primary inferences are validity of purpose, ethical aspects and likely consequences.

That approach reaches into much deeper channels than those limited perfunctory analytical methods practiced by too many science and technology analysts. Now, this is not to infer that such an approach will yield perfect answers in an increasingly complex world. Rather it is to underscore that all professions should enhance their awareness of a universally applicable method which places metaphysical inferences—or, if one prefers, intuitionally perceived inferences—in a decisive role. This, in turn, would bring the various professions and their respective specialisms more incisively into sound interdisciplinary endeavor. Thus, their efforts would yield nearest right answers to complex issues, or at least a posing of the right questions for more focused study.

Popular critiques on technology and the profit motive versus ecosystem and socio-economic decay will also be addressed. However, the design of the OMEGA matrix, touching as it does on all professions and factions and the resultant world ecosystem, is addressed to the highest possible order of things—constitutional principle in the context of those inferences leading to the optimum ethical parameters of humane purpose.

In this ultrastructure of things, metaphysical and intuitional inferences will be addressed throughout the text as the first order of investigation. However, these inferences will not be dissertations of a theological nature, though often so inspired. Metaphysical inferences can also involve quite earthy matters.

For example, our industrial technology can produce a large economic growth gain on one hand, while deeply poisoning a major waterway or water table on the other. That people should live without harassment in a poison free environment is a metaphysically circumscribed ethical and moral tenet of life. Uprooting suburban neighborhoods, following long term negative effects of industrial poisoning, can only be described as a result of gross indifference to moral and ethical parameters on the part of the perpetrators.

Integral to the philosophy of the OMEGA thesis is that while there are myriad professions, religious beliefs and factions in the melting pot of the world economy, all of the players in the qlobal enterprise fall into either of only two categories—those who respond to constitutionally inspired ethical parameters and those who do not.

Parenthetically, while the generalization still holds, there are regretable gray areas here. Of course there are those who are overtly antithetical, and for them no gray areas exist. On the other hand, there are those highly motivated endorsers of ethical and moral principles who all too often find themselves in a gray area between a rock and a hard place.

Consider the farmer adjacent to a major metropolitan area who is forced to yield his one-thousand acre farm to industrial and urban sprawl. This particular farmer is a graduate of a top agricultural college, he is a dedicated environmentalist and is well informed on the parameters of economic growth. In his particular situation he knows that the industrial plans for his thousand acres simply can not be contained within the existing already critically overstressed metropolitan infrastructure. Other than his participation in a like-minded local citizen group, who lost their case to the politically motivated environmental charlatans, he has no other choice. Not having the resources to wage further battle, he joins the more indifferent farmers adjacent to his property and reluctantly yields to forces he can not control...and moves far away.

Underscoring all this, notwithstanding land use zoning changes which **are** legitimate, there are many more zoning changes made within existing overstressed infrastructures which are not containable.

Gray areas abound here. The concept of geographically dispersed great new cities with new core infrastructures—new metropolitan centers which could properly accommodate seemingly endless exponential population growth—has not yet persuaded governing powers. This leaves a political mish-mash which often triggers unsound compromises in public utilities and ecosystem control by well intentioned people who, as victims of their administrative gray areas, can only try their very best to "make do".

Hence, maximum clarity is sought after in voicing those conditions precedent to looking away from antithetical control and toward One God for guidance—guidance rooted in constitutional principle in steering the global enterprise toward a coherent ecosystem.

But at the present time, inferences and ideas attending the access to physical parity for all people, though beautifully exhorted by astute investigators, are largely either dormant within institutional research libraries or found stagnating in response systems which are not holistically structured and which lack the political clout to bring about warranted change. In short, we do not respond in any meaningful degree to the many excellent investigations of metaphysical inferences in the daily operational mainstream of economic growth. We lack a dynamic platform for placing issues in the context of the whole of things. As touched on earlier, the constitutionally inspired minority continues outvoted to an alarming degree. There exists a desperate need for access to some grand synthesis of the whole of things common to all who ought to participate in the decision making process.

Lacking such a platform, the antithetical Philistines have been allowed to take over, their mechanistic credos seeking only quantitative economic growth. They hardly question the clear and present danger of a technology which has become a headless monster, lunging forward under its own unbridled momentum, leaving physical and spiritual devastation in its wake…In a word: Techno-Anarchy!

The central advocacy of OMEGA is founded on the ethical parameters stemming from constitutional principle…not on unbridled Techno-Anarchy. Responding, then, through this principle to the ever widening array of forces confronting us in the twenty-first century, OMEGA is an attempt to offer response to highest order practical purpose—that **one** aspect of the whole of things which subsumes all other factors. That one aspect is to contain the "Ecosystem Countdown", described in detail in the next chapter. In short, OMEGA will be offered as a concept which yields a macro-methodology for steering the whole of global enterprise to a coherent ecosystem.

Idealistic?…Yes, but still practical in its plea for educational and operational processes which can be implemented over a digestible period of time.

Toward closing this opening chapter, the metaphysical principle holds that **man does not create things**. Indeed, **man discovers things** which in God's Eternity have always existed as ideas or natural laws. As discoverer then, not creator, man should seek to discover that grand synthesis of natural laws and ideas which can absorb individual discoveries in their proper relationship to one another and in their relationship to the whole of things.

Parenthetically, it should be noted that an astounding number of professionals shrink away from the term "metaphysical" when approached in the work environment. Even among the churchgoers, it seems that when they leave their respective churches on Sunday after listening to and even endorsing metaphysical exhortations, they do not make a potential connection with professional problem solving. Such thinking all too often stops at the church exit. With too rare exception, metaphysical precepts and secular professionalism are not blended—particularly in science and technology. A vast number of professionals are simply not listening to the great scientists who openly encourage response to metaphysical precepts in solving secular problems. The great scientists and those who listen to them appreciate that the quest for a better world rests upon two pillars— "The pillar of religion" and "The pillar of science" in the context of "world loyalty" as introduced in the opening of the prologue.

However, one can observe that those who resist the term "metaphysical" in professional activity can often be reached through the term "intuition". Be reminded that intuition is a metaphysical derivative, one of God's greatest gifts to each and everyone. All people possess it. All have experienced it. Experienced in varying degrees, yes, many times not even consciously aware of having responded to intuition. But intuition can be nurtured and consciously expanded by anyone. The rewards are profoundly gratifying. (A must-read book: Ellie Nadel's "Sixth Sense" for penetrating discussion of intuition and related "whole brain" research.)

In these closing paragraphs of Chapter One, it is also appropriate to acknowledge that in addressing the global ecosystem problem in the context of America's constitutional principle, we must deal as best we can with opposing beliefs and practices, not only in America but in other countries as well. There are vast domains which are not only dominantly atheistic or agnostic, but which overtly seek to spread their beliefs to other areas. One of the more gross examples of this is China's plan to actually teach atheism to Tibet, suppressing a nation premised on spiritual perception.

Obviously, as leader of the free world, America's approach to a world ecosystem based on the ethics of the human condition will be even more difficult than within our own borders. But pursue it we must since it is equally obvious that the entire globe is affected. The total ecosystem is now without boundaries. Nonetheless, perhaps it is not too unrealistic to suggest that ethical parameters can eventually be sold to governments like China on the raw, fundamental basis of global survival and the assuaging of hunger.

It is here, then, that without abandoning U.S. constitutional principle, we can reduce tendencies toward religious argument by simply holding to those ethical parameters which, once derived, stand on their own without further reference to religious precepts. This means that in so approaching the now borderless global ecosystem problem, we are essentially freed from engaging in world-wide contests among theists, atheists and agnostics. That as well as the seemingly unending bitter internecine wars being fought between the sects of Christianity, sects within Islam, sects within Judaism, sects within Hinduism, sects within Buddhism.

Now, this is not to deny the ultimate possibility of reaching wider global purview of America's constitutional platform, but it is not mandatory at this point in history. The words of Alexander Hamiltion express with poetic beauty the path of mankind toward an inevitable peace, with physical and spiritual equity for all. Hamilton said:

"The sacred rights of mankind are written as with a sunbeam, on the whole volume of nature, by the Hand of Divinity itself, and can never be erased or obscured by mortal power."

To these good ends, then, it is hoped that all professional disciplines will one day discover in the OMEGA matrix the nucleus of a macro-system which offers a basic paradigm for a sound approach to humane purpose. In the OMEGA matrix, an **ethically manageable** home for all issues can be found—all issues relevant to proper operation of the global enterprise, moving us away from inhumane ecosystem chaos toward humane ecosystem stability.

*   *   *

CHAPTER TWO

The Ecosystem Countdown

It is vital to the OMEGA thesis that the dynamics of history be portrayed in such a manner as to sift out nonrelevant lower orders of history. In turn, it is necessary to show the dynamic interconnectedness and the teleology of history's epochal events. History is not Zeno's arrow, piercing static snapshots over time and calling it history. History is comprised of a shrinking time dimension, an ever increasing tempo of life and a teleology of ultimate Divine purpose which includes the ethics of conduct toward realization of that purpose. Yes, history is a time-tempo-teleology interlock and has much to teach us which is not yet sufficiently embraced by the education system.

The above dynamic constituents of history, not having been previously reckoned with, now find us in an increasing rate of uncontained overlapping landmark events. This has been brought about largely by unbridled technology. Some of this is good, some of it bad, and all unmanaged when measured in the context of the critical and growing instability of the world ecosystem.

Hence, a big screen portrayal of history is offered, contrasting the explosive uniqueness of the twentieth century with all that preceded it during the earlier, then-digestible evolution of man's tenure on earth. Such big screen perspective tends to make more obvious the little understood phenomenon of the time-tempo-teleology dynamic. This will condition our thinking for grappling with time compression.

However. at the outset of this chapter it must be underscored that the concept of time compression, and the ecosystem countdown it has forced upon us, is too easily waved off by many professionals. Yes, all can easily agree that technological growth is outpacing our ability to properly and safely manage it, and that this problem has been increasing exponentially over the centuries. But having easily agreed with that oversimplification, the general noncorrective behavior of the parties responsible proceeds unabated.

The pile up of those events which restrict or often even stop normal life-flow systems is on a sharp increase, notwithstanding some piecemeal efforts to address the problem. The general indifferent outlook to the effect that "somehow it will all work out" does not make for containment of the ecosystem countdown.

In lecture work, the writer has consistently encountered a need to move beyond easily acknowledged perfunctory statements about technology's growth outpacing our ability to manage it. Yes, there is a need to embrace a big screen portrayal of time and tempo, and, moreover, a need to account for the evolutionary undercurrents that have subtley given birth to that now uncontained twenty-first century dilemma. Hence, a comprehensive but summarized view of the time-tempo-teleology dynamic is offered.

The great curve shown on Figure One was initiated the day the first primordial vertabra crawled onto the shores of the Paleozoic Sea. But we need not go back 500 million years.

From the standpoint of the OMEGA thesis, our entire interest lies within a cosmic speck of time. The great curve, though eons in the making, holds significance for is only in terms of recent millennia. In fact , the great curve barely hinted of geometric progression until the epochal event of Columbus opened up the global seafaring era only five centuries ago. Even then, the phenomenon of time compression had not been observed on the distant horizon.

With big screen perspective, then, let us examine the whole curve with the thought of **really feeling** the severe contrast between "then" and "now". We shall underscore how time compression and attending ecosystem instability seems to have caught our leaders and technology unaware of the magnitude of the "time war". Until reckoned with, that war will militate against world loyalty and humane purpose.

The ecosystem countdown, then, within the cosmic speck of time we shall address, begins with the "tenth second" in our "ten second" countdown. Everyone, of course, is familiar with the minus-ten-seconds-to-launch countdown procedure in a space vehicle launching—minus 10 seconds-9-8-7-6-5-4-3-2-1-0. Launch. A useful perspective obtains in finding the global enterprise as a whole in a cosmic "final ten seconds" prior to launching the earth as a viable, operationally ready entity in the Creator's grand scheme of things. See Figure 1.

WORLD POPULATION GROWTH AND KEY EPOCHAL PERIODS
TIME COMPRESSION
and
THE ECOSYSTEM COUNTDOWN

The teleology of this curve, as viewed with highest order epochal periods superimposed, reflects the essence of history's dynamic…inferring the "time compression" of bvents and the attending spillover of ecosystem decay.

10. AGRICULTURAL AGE

9. ERA OF PHILOSOPHICAL AND THEOLOGICAL SYNTHESIS

8. GLOBAL SEAFARING ERA

7. INDUSTRIAL AGE & LANDMARK CONSTITUTION—"IN GOD WE TRUST"

6. ELECTRIFIED COMMUNICATION ERA

5. AEROSPACE ERA

4. NUCLEAR ENERGY ERA

3. MICRO-TECHNOLOGY & THE INFORMATION ERA

2. SUB-MICRO-TECHNOLOGY ERA—MOLECULAR & GENETIC SCIENCES

1. ENERGY CONVERSION VERSUS'ECOSYSTEM CLIMACTERIC

0. MOVEMENT TOWARD A BALANCED GEOSPHERE-BIOS-PHERE-SOCIOSPHERE

"Minus Ten Seconds"

So, the agricultural age, looking back some ten-thousand years, seems a logical and practical starting point from which to address the "tenth second" of our ecosystem countdown. At that time the entire population of the earth was estimated at only five-million people. But geometric progression would now be sharply accelerated, triggered by agriculture. This made for the ability to feed more people and the gradual abandonment of the herder-gatherer life in favor of fixed locations. In turn, this led to the development of towns and cities, sewing the seeds of an ultimately runaway population growth. Within a few millennia the agricultural age saw large scale civilizations rise and flourish with rapid cultural evolution. Even then a kind of political structure emerged which, in some respects, persists today. And so the early millennia of the agricultural age proceeded without further events of great moment, in the context of the big screen perspective, that is, to Year One A.D.

"Minus Nine Seconds"

Year One marks the median point on the era of philosophical and theological synthesis. This "ninth second" of our countdown embraces an analytically useful time on the great curve for a synthesis of thought which flowed across several centuries. In a few "micro-seconds" of the "ninth second", we sweep from the completion of the Parthenon over four centuries earlier to the great integrative thinking of Marcus Aurelius in the second century A.D. This period, of course, embraced the early wisdom of Socrates, Plato and Aristotle, The Old Testament,

The New Testament and the best thinking of the Roman Empire. Altogether, this body of thought is very much alive in man's view of things today.

"Minus Eight Seconds"

And so to the global seafaring era, the "eighth second" of our ecosystem countdown. Now, bearing in mind that we are attempting to develop a clear view of the heretofore unseen but always present teleology of events leading ultimately to time compression, note the trends thus far. We already see that where eight-thousand years passed between the "tenth second", the beginning of the agricultural age, and the "ninth second", Year One, only fifteen-hundred years pass without events of great moment between Year One and the "eighth second", the global seafaring era. Again, in the context of the big screen perspective. Crossing Year One with a population increase to two-hundred million people, we reached the global seafaring era with a doubling to four-hundred million people.

Parenthetically, it is noted that many pre-Columbian scientists and philosophers knew the earth was round. Even Aristotle knew this, but for him the earth was an inert bauble hanging in the heavens. The scientists did not get their whole act together on the mechanics of the solar system until around the time Columbus **operationally** demonstrated the round earth for all to see. Then, with relative suddenness, world wide commerce emerged since mankind at large—thanks to Columbus, Vespucci and Copernicus—now knew of a rotating, sun-orbiting, round earth.

That epochal period of the "eighth second" in our countdown yielded an imposing array of new discoveries, actions and ideas of great moment, and also presented much teleological evidence of predestiny. Not the least of this was that of moving man's thought away from the myth of a flat earth at that very point in his evolution which found him ready for a quantum jump in complexity.

A few decades to either side of the initiation of the global seafaring era, a climacteric of fermenting thought was being resolved—for the time being, at least—as great minds were further refining the scrolls of early scripture, refining the best of early Greek and Roman concepts of government and jurisprudence, synthesizing the best of art and architecture with the scrolls of philosophy and science and so forth. In general, though perhaps not consciously aware of it, the great minds were setting the stage for the Renaissance of western civilization.

But great thinking must now be promulgated beyond the extremely narrow boundaries of communication permitted by hand written scrolls. So, enter Gutenberg on teleological cue. The machine printed word, appearing in the middle of the fifteenth century, within a few decades began mass production of the publication of great thought. It is teleologically noteworthy that man's ability to disseminate great thought on a broad base did not evidence itself until great thought could demonstrate sound points of departure. Most notably, of course, was the round earth and its award of intellectual freedom from the gross myths attending the concept of a flat earth. Moreover, the printed word, as evidenced in its first glorification through the Gutenberg Bible, was destined to serve the Creator's grand scheme.

The Renaissance of western civilization was now in process on the seaways of a round earth. As the spread of the printed word engendered the body intellectual and the body politic, all men could participate in intellectual and political processes to a degree of their choosing. The Bible moved into the home. Libraries of great literature emerged. The tempo of life was markedly on the increase. The enterprises of men were coalescing, triggering new needs, new ideas, bold actions.

Parenthetically, it is interesting to note that within a few "microseconds" of Columbus setting foot on the sands of a now affirmed round earth, one of Roger Bacon's rebellious admonitions took hold. Bacon, in one of his disputes with some poor translations of Aristotle, was attempting to ready mankind's thought for the inevitable Renaissance:

"Cease to be ruled by outworn dogmas and authorities", he shouted. **"Look at the world."**

While Bacon did not live in the century that saw meaningful implementation of his grand admonition, the scientific luminaries of Columbus' time did, indeed, "look at the world". Led by Leonardo Da Vinci, probably the greatest of all Renaissance men, followed by others over the next couple of centuries—including Copernicus, Descartes, Spinoza, Galileo, Issac Newton—modern science was platformed.

"Minus Seven Seconds"

The stage for the "seventh second" was set. So, little more than two-and-a-half centuries after Columbus, the industrial age was greeted by a population having almost doubled still again to over seven-hundred-million people. The "seventh second" of our ecosystem countdown bore witness to one of history's earliest great concurrencies—the industrial age, the theologically premised Constitution of the United States and the then platformed modern science.

Unlike the impact of the agricultural age, localized as it was to small populations on a "flat earth", the industrial age marked an enormous impact on the way people would manage their round earth. An exponentially increasing population growth was now an acknowledged fact.

But above virtually all else, the intense conversion of energy was now to prevail throughout all time. Energy was soon to become the backbone of an explosive increase in the tempo of man's enterprises. With the invention of the steam engine and the ever increasing use of machinery, not only a quantum jump in energy demand took place, but energy demands quickened exponentially. Now a two-headed monster manifested itself. That is, fossil fuel was required, of course, to run the machines which produced subsistence products for people. But! Not properly acknowledged in the demand equation was that fossil fuel was also required to run the machines to acquire still more fossil fuel to run the machines to produce still more products for people. Moreover, the two-headed monster breathed the fires of extremely high entropy. Spewed wastes accruing to that process were virtually ignored. Energy demand and attending entropy effects would prove to be a never-to-end, fiercely self compounding, ecosystem quality problem.

Parenthetically, in any context of population growth versus subsistence, we must pause to note that Malthus' controversial thesis on geometric population growth versus arithmetic subsistence acquisition was certainly not wrong insofar as the spirit of his warning was concerned. Indeed, his warning to the effect that geometric population growth was incompatible with arithmetic subsistence growth might well have set in motion the trends which would eventually place subsistence acquisition on a corresponding geometric curve. Particularly so in the light of the twenty-first century with a world population growth to ten-billion people on the mid-century horizon. To be sure, Malthus' early warning reemerges. Reverend Malthus' original thesis, appearing in England concurrently with the industrial age, set the tone of a vitally important branch of new thought. No demographer enters into speculative thought on population growth trends without due homage to Malthus. Moreover, in a teleological sense, it is not unreasonable to suggest that the seeds of Malthusian thought were sewn precisely at the right moment in history for proper incubation.

"Minus Six Seconds"

And so we move from the opening of the industrial age and in less than a century we cross the one-billion people mark to the "sixth second" of our countdown—the electrified communication era.

Again, the teleological inference. The global enterprise was now at a point in time where it was becoming clogged in communication delays. So, enter Morse and Marconi, emerging from ivory tower like isolation, to give us the capability to communicate at virtually the speed of light precisely when it was most needed…Time compression was now felt both individually and collectively in the world community, but in the middle of the nineteenth century time compression was still easily contained.

"Minus Five Seconds"

And so to the dawn of the twentieth century and the momentous "fifth second" of our ecological countdown—the aerospace era. Population was now only a few decades away from the two-billion people level.

In the light of the initially unseen magnitude of the impact of powered flight on the affairs of mankind—land ways and sea ways now to be augmented by air ways—it is noteworthy that this century opened with great contemporaries of the calibre of Einstein, Russell, Whitehead, Bohr, Bergson, Huxley, Oppenheimer and others. The modern science platform established just prior to the dawn of the industrial age would now be updated with highest order syntheses of scientific and philosophical laws, axioms and maxims.

The efforts of these early twentieth century luminaries, concurrent with the advent of unsophisticated but portentous powered flight, opened pathways to the acceleration of new discovery. Moreover, they proffered ways of thinking whereby new discovery could be integrated with existing modes.

The vital phenomenon of time compression, however, while latent in the grand scheme of things, and while obliquely inferred by Henri Bergson, went essentially unanticipated. Analysts were, perhaps, caught up in the wonder of it all. Who would ever have visualized that what emerged from the Wright Brothers' humble bicycle shop would evolve so quickly and have such impact on the affairs of man?...Who could have imagined, in a few short decades, ending a great war with a high flying long range airplane capable of carrying a nuclear bomb?...Who could have imagined a supersonic airplane taking off from London at breakfast time, racing the sun, and having breakfast at the same time in New York?...Or, indeed, who would have imagined flying to the moon?

In any case, the advent of powered flight at the turn of the twentieth century found that the great curve would now, in the brief time span of less than a century, explode near vertically in a chain-reaction-like concurrency of human births and new technologies—nuclear energy, micro-technology and sub-micro-technology, etc. This rapidity of growth would compress time beyond the limits of man's capacity for efficient event containment under existing system design philosophies. Emphatically so, since human population was by then approaching five-billion people, now doubling in billions rather than mere millions.

Since history by then was more clearly exponential than ever, observe for a moment the degree to which this explosive growth situation suffered a lack of response to the emergence of time compression. Note that as the decades of the second half of the twentieth century passed, a severe pile up of incorrectly handled operational events was occurring with a severe increase in the rate of system malfunctions, failures and, most disconcerting of all, human error. This, not so simply, was a matter of system designers failing to provide operationally coherent designs which would not overload the human operators in a climate of ever increasing loads on the system. These time compression loads increased as larger and larger numbers of people to be served were fed into systems which were inadequate to begin with.

The net result of all this has been the incipient deterioration of the global ecosystem. This was not seen for many years and even today acknowledged in only a very limited way. Our operational systems appear to function, albeit very inefficiently, but they crudely serve their purpose…and, for the most part, continue to crudely serve their purpose as ecosystem concerns deepen.

For now, having covered a few "micro-seconds" of the "minus five seconds" of our ecosystem countdown, the aerospace era is briefly summarized. The most pivotal event in all of the Great Curve's history was the birth of the aerospace era.

Why the most pivotal event?

The more obvious point, of course, is the breaking away from ponderous surface travel as the only way of moving about and interfacing the world. In a mere few decades out of the Wright Brothers' bicycle shop, we were viewing and interfacing the entire world at great heights and distances, setting the stage for an explosion in the scientific, technological and industrial affairs of man. Most notable was the deep and vigorous plunge into the microcosm of atomic physics, motivated and accelerated by the availability of a flying machine capable of carrying a war-ending nuclear bomb to its destination. The aerospace industry's role in the acceleration of scientific achievement and world economic growth thus far is only a beginning. Its further role as a major participant in resolving global ecosystem management will be covered later.

"Minus Four Seconds"

The nuclear energy era, at its beginning, had no sooner witnessed the termination of World War II when the pursuit of nuclear energy for peaceful purposes was launched. Sadly, this effort was accompanied by further pursuit of even more vicious nuclear weapons. Too, there is a deep dilemma in the use of nuclear power for peaceful purposes wherein it is the only "clean" energy source available in great quantity at this time. This, notwithstanding the operational dangers of aging equipment and the yet unresolved massive waste disposal problem.

That situation, and the time scale to suitably massive applications of dynamic based energy sources—solar, wind, hydro, geothermal—together with a substantial phase out of ecosystem-devastating fossil fuel energy sources,—is certainly one of the most urgent problems facing the macro-management of science, technology and the politics of realignment of industrial goals.

All in all, the ever deepening scientific exploration of the microcosm was producing good things and, only two decades beyond Hiroshima, found research and development activity well into the micro-technology of information processing. Also, quantum physics was brought into a prominent role in charting the course of scientific and technological development…More later on this.

"Minus Three Seconds"

So, we approach the "third second" of our ecosystem count-down-micro-technology and the information era. Even at the outset, the leading question addressing the intense computerization of information process-ing was not "What are our real information needs?", but rather: How small can we make the physical system, how much information can we cram through the system, and how fast?"

It is acknowledged, of course, that on the positive side, "how small", "how much" and "how fast" has served society well in hard data infor-mation areas such as banking, accounts processing, health care, etc. The speed and accuracy with which enormous numbers of people can be served in areas vital to the orderly management of their personal lives is a good thing. This is an all too infrequent but highly commendable example of reckoning positively with time compression in our explosive economic growth situation.

On the worrysome side, however, is the fact that the increase in the ability to process more information at a faster rate is adding miserably to the further proliferation of nonrelevant information glut.

Moreover, we must underscore a major caveat bearing on the imme-diate future of information processing. That caveat is our failure to reckon with the consequences of placing intensely computerized devices at the disposal of the young minds who will inherit the manage-ment of the world and its global ecosystem. These young minds are not being properly educated to the real world time flow of the operational systems through which we live our daily lives. The natural time rhythms of nature—before which every technological head must one day bow—are virtually in a process of abandonment to the unnatural time struc-tures of information processing technology.

Information computerization as a servant to contemporary mankind is a must, of course. But we are allowing young minds—the future rulers—to become the slaves of nanosecond computerism, dealing in micro-fractions of a second, and engendering a still greater drift away from a principled response to the natural time rhythms of nature. It is underscored that we must constrain the central thrust of nanosecond computerism in its push for deeper advances in the state-of-the-art within the microcosm of nanosecond technology per se…That is, until technology at large pauses for assessment of operational legitimacy in the time flow of natural rhythms in the real world.

It is the above noted situation, combined with fragmented soft data pouring in from every conceivable source which poses such fierce problems in information management—being under constant pressure to sift the glut of nonrelevant information in often futile attempts to find the relevant. Even in the computer field itself, the humorous adage as to the worth of any computer program persists in jest rather than admonition—"garbage in—garbage out". And so the outreach of micro-technology toward ever smaller and faster information system computers with ever-increasing capacities for more and more unstructured information "garbage", until constrained to legitimate needs, continues unabated.

On the positive side, however, we noted the airplane as the catalyst which triggered the deep and vigorous plunge into the microcosm of atomic physics. In the light of that, we probably gained some decades in making the needed discoveries waiting for us in the microcosm of things, carrying us rapidly into micro-technology, and, as we gained still another billion people, on into the sub-microtechnology era. All that is good, of course, only insofar as caveats are reckoned with and this increased knowledge put to legitimate use.

Teleologically, the intense need to bring forward the riches of the microcosm in order to grapple more effectively with the needs of several billion people was predestined. On the other hand, we must observe grave miscarriages in operational utility along the way. Now, the argument was never whether or not we needed the riches of the microcosm. The argument is addressed to severe but avoidable entropic stumbling as we move toward our predestination. The crux of it all is that in our headlong pursuit of the microcosm we have all but lost sight of the enormous scale of the macro-economy now upon us and yet to come, together with its attending grossly deficit ecosystem. The "micro" of physical science is in critical imbalance with the "macro" of operational systems utility...This aspect enlarged later.

"Minus Two Seconds"

So, we arrive at the "second second" of our countdown, the sub-microtechnology era—molecular and genetic sciences. But not yet having a macro-management structure of the whole of the global enterprise, we cross the five-billion people level without a steering mechanism for guiding the now massive microcosm of sub-micro-technology.

Yet we note magnificent things happening, too. The genetic engineering of agriculture, for example, laboratory crafted seeds and plants of much greater efficiency. One can observe in this a none too soon development as we cross the six-billion people level with an ever increasing need to feed more people. This, of course, applies not only to the hungry people in the world today, but to an additional four-billion people over the next half-century.

A caveat, however, is found on the other side of the molecular and genetic sciences coin. We see rampant research incursion into the human aspects of genetic research. Much of this is to be highly appreciated as sound growth in the state-of-the-art of medicine. On the other hand, we must be alert to an invasion of the precepts of Divine Causality as science and technology seriously consider the cloning of human beings.

As reiterated or inferred in various ways throughout the discussion thus far, mankind stumbles painfully along the way at times, invoking discoveries which, on one hand, benefit all mankind and, on the other hand, suffering the wars and unethical byways of the very technological catalysts which brought the good things to pass in the first place. Nonetheless, with proper focus on the global ecosystem, there is every probability that as we enter the twenty-first century with six-billion souls on earth, sub micro-technology will join the other players in the global enterprise and come forward with many answers to the need for advanced subsistence processes...demand and supply one day meeting in the teleology of self-fulfilling prophecy.

"Minus One Second"

And so to the last second in our ecosystem countdown, the energy versus ecosystem climacteric. As we have noted, the energy axis— energy converted plus entropy rate equals ecosystem status—is the backbone or major axis of all that takes place in the promulgation of the global enterprise. In this "last second" of our countdown, we enter the twenty-first century in a climacteric of noncontrol of the excesses of energy conversion.

This is not to say that good things are not happening in the direction of corrective action. Literally hundreds of organizations now exist for the sole purpose of promoting ecosystem advocacies of one kind or another. These organizations reach from grass roots level all the way up to important new congressional committees and top government administrative functions—all aimed toward reckoning with some aspect or another of the energy axis climacteric now upon us.

But as brought out in various ways in our discussion thus far, most of that effort endures without a structured approach to efforts which can only be described as chaotic. Too, that effort proceeds without general acknowledgement of the severity of time compression. Hence, in the remaining "micro-seconds" of the "last second" of our ecosystem count-down, much is yet to be done before we reach "zero seconds" and finally "push the launch button" on an operationally viable Space Ship Earth.

"Zero Seconds"

In the remaining "micro-seconds" of the "last second" of our ecosys-tem countdown—movement toward a balanced geosphere, biosphere, sociosphere—the synthesis of structured actions implicit in the OMEGA matrix can be brought to bear...hopefully before moving too far into the twenty-first century. Now, this by no means suggests the optimum situation that soon. As earlier noted, OMEGA is designed to mature over time. But, uniquely, the elements of OMEGA will tend to coalesce in the context of the whole design.

OMEGA is not expected to come into play all at once; but rather to place myriad disjointed fragments of our economic growth and ecosystem balancing efforts in their proper and **clearly observable** relationship to one another. At least wide understanding of the nature and magnitude of the ecosystem problem can be achieved. This means that an induced prioritization and progressive integration of fragmented efforts will be one of the greatest initial gains to be realized through the OMEGA matrix.

It is not too unreasonable to suggest that within a few decades, all of the constructive ecosystem efforts now working without a central structured aegis, without a viable relationship between the industrialist and the environmentalist and without a sensitivity to time compression, can be brought into full concert with one another. This would lead, eventually, to a balanced global ecosystem and to "zero seconds"—launch!...

In closing Chapter Two, it is appreciated that a book length treatment of the gceat curve alone might be in order. But the whole point here is to deal with a major dimension of principle—a specifically oriented macro-focus on history's dynamic, the "ratchet points", if you will, on the great curve...and not myriad lower order events of history which help move the ecosystem countdown from one ratchet point to another.

Most informed readers will respond to the ten epochal cues and fill in for themselves anyway. For example, the global seafaring era. The fact that Columbus refuted the flat earth concept and locked man's tenure on earth at a higher ratchet point, with all that is implied in that, is quite enough to satisfy our rationale. How much or how little Aristotle knew of his round earth observation is not important   to our thesis. Nor is Galileo's hassle with the church as to the mechanics of the solar system of any significance to our thesis...What is significant is that Columbus' **operational** affirmation of the round earth ratcheted all of mankind to a higher set point as to the intensity with which the global enterprise was to be conducted.

This viewpoint applies to all ten epochal ratchet points. The whole intent here is to illustrate, with as little lower order history as possible, the true time compressing dynamic of history over a ten-thousand year period. Portraying this history over that span of time is intended to place contrasting emphasis on the severity of today's rate of evolution, and thereby encourage greater attention to exponential thinking in the analysis of economic growth and ecosystem status.

So there we have it. The ecosystem countdown via the great curve, reflecting the time-tempo-teleology interlock and the inferred compression of time.

*   *   *

CHAPTER THREE

Summary of Issues, Principles, Motivations

In the prologue, we outlined the pivotal issues which have placed the world in a deteriorating ecosystem and a now rapidly diminishing quality of life. In summary, these issues are:

- Religion versus science not yet reconciled.
- Natural and man-made environments clashing.
- Entropy law ignored—ecosystem paradigm nonexistent.
- Economic theory unable to reckon with today's needs.
- "Time compression" largely ignored—pile-up of events.

In turn, we introduced certain of the basic factors to be dealt with in overcoming the above obstacles to humane economic growth. Required actions are implicit in the discussion of constitutional origins in Chapter One and the time compression dynamics of the ecosystem countdown in Chapter Two.

In this chapter, and before further enlargement of the above in subsequent chapters, we shall summarize the basic motivations of certain of the players in our ecosystem dilemma. This will not be a clinical discussion. We'll leave that to the psychologists. We will simply offer a brief, down to earth set of observations on those behaviors which inhibit proper regard for the core ethical reality of things as they bear on reckoning with ecosystem deterioration and the human condition at large.

But first, let us briefly summarize the basic principles discussed in the first two chapters:

- **The principle of constitutional ethics,** which in summary of the Chapter One discussion, holds that only two kinds of people inhabit this world—active constitutional advocates and those who engage in antithetical practices; and that basic reconciliation of the two, or movement toward that end, is to rationalize all problems associated with the ecosystem and related humane purpose in the context of constitutional principle and ethical parameters…gray areas notwithstanding.
- From this highest order principle, then, the subordinate principles reckon with the operational dynamics of history in the context of ever-accelerating economic growth. This also includes reckoning with those key issues which explain the world's current state of ecosystem chaos. The subordinate principles are:
- **The principle of time aompression in exponential history,** which holds that the dynamics of history are reflected in those highest order epochal periods or events which have compounded to still more events in geometric progression across the centuries…until the second half of the twentieth century manifested itself in the form of severe time compression. This means that the developers of the world's operational systems—not having anticipated and designed for time compression—have allowed these systems to grow and overlap one another to a point where they are faltering or failing for lack of linear time to contain all of the actions necessary to maintain necessary order and, in turn, to inhibit ecosystem decay.
- **The principle of the ecological evolution of man,** which holds that generic man is in a continuing state of evolution in his relation to his spiritual, mental and physical capacity to contain his time-compressed life style. This evolution of man, in its proper sense, means that man must be allowed evolutionary

increases in his **naturally endowed** capacities. Necessary further augmentation of man's abilities must be devised as independent machines and processes separate from his mind and body. More later on endorsement of medical advances on one hand, while on the other hand firmly rejecting genetic engineering for humans or, indeed, rejecting any practice which seeks to "rewire" human beings for accelerated mental functioning or increased physical dexterity. In short, solve the problem with more responsible design.

- **The principle of determinant teleology,** which holds that final causes exist and that there is evidence of design and purpose in all phenomena, and that further evidence exists to the effect that phenomena move toward goals of self realization—and that, ultimately, this is good. This means that, among other things, man has the option of observing the behavior of current events, historical epochs and antecedents as they coalesce to suggest more accurate formulations of the future. In turn, man can then structure his present actions to better phase with that approximately knowable future…However, in the context of teleology, one must be keenly aware of the un-God-like force of dysteleology. This is the doctrine of purposelessness in all nature which, though perhaps practiced unwittingly in some cases, is nonetheless an antithetical force of some consequence very much at work in the ecosystem climacteric now upon us.

In summary, then, when we look to the principle of constitutional ethics and key subordinate principles—

- Time compression in exponential history (pile up Of events)
- The ecological evolution of man (capacity for increased tempo)
- Determinant teleology (nature declaring itself)—

We have a time-tempo-teleology interlock which platforms our approach to a global ecosystem governed by humane purpose.

At this point, a few comments are in order on the motivations of some of the key players in the global enterprise. The following are summary highlights, some of which will be enlarged in subsequent chapters.

Consider, for one key player, the motivation of the scientist or technologist, who upon graduation from school starts his career as a technical specialist. Through his technical brilliance he makes his way into the upper levels of management. Since he has no direct field experience in the real world of operational systems, he has a psyche quite different from that of the very operational system client to whom he is selling his product. The specialist is heavily steeped in perspective-limiting quantitative factors. Hence, he all too often evidences insufficient regard for those qualitative real world operational factors which can often bend a major design decision in a sharply different direction.

Now, this does not for a moment suggest the brilliant specialist as irresponsible. What is being suggested here is that he simply does not know what he does not know. Much of this, of course, is to be laid at the feet of the educators. The educators, at this period of history, are still very weak in the scope and intensity of their interdisciplinary and generalist training. Sharply increasing the population of technological generalists, placing that population in balance with the now overwhelming population of specialists, is vital to immediate and long term reckoning with our deteriorating ecosystem…More later on this critical factor.

Before continuing our critique of the players, it bears repeating to note that among the scientists, technologists and industrialists, some good things are happening with regard to reckoning with ecosystem deficiencies. But those noteworthy efforts border on the insignificant when laid against the propensity of most industrialists to place the profit motive above all else. It is simply not yet widely appreciated that, long term wise, the making of profit is not incompatible with achieving a sound global ecosystem. In the meantime, consider briefly the current way of things.

The brilliant technological specialist, now in the upper management levels of industrial enterprise, is not only steeped in a microcosmic view of things, but is often constrained to a kind of tunnel vision inherent in preoccupation with the profit motive. No matter that the energy-conversion-versus-ecosystem-climacteric is the result of the now rapid deterioration of our ecosystem on a global scale, the technologists, real world problems notwithstanding, are very aggressive in their public image building. The universities, the Rotary and Kiwanas clubs and all other public forums are awash in passionate speeches about the great future to be brought about by science and technology. These speeches by the "corporate seers" from the technology community lure far too many uninitiated minds afield of the real problem through the seer's urgings that the public find technology in good hands.

It has long been fashionable for scientifically steeped corporate officers to charge briskly about the country as prophets and seers of our technological future. Unlike the environmental scientists outside the profit oriented corporate structure, the corporate scientist, representing the vast majority of technological influence, makes his statement as mere projection of today's scientific advances, generally without regard for the great overriding issue of an increasingly inhumane global ecosystem.

In any case, it is the public reaction to the corporate seer's speech that should concern us. The general reaction to thousands of talks on the bright future promised by technology is that technology will take care of us and give us more of the good life. We not only want to believe this, indeed it is possible...once the lip service to the greater issues,which often close such speeches,becomes the new corporate reality. In the meantime, the corporate seer returns to the immediate pressures of short term profit thinking...Future deferred until the next invitation to make a speech about the future. As a servant of the profit oriented structure of the industrial organization, the corporate scientist's speech about the future, with too few exceptions, can be permitted only as a showpiece device for building the public image of the corporation.

As for a motivational critique on domestic policies and geopolitical practices, this near-impenetrable milieu will be folded into subsequent chapters as one of the more difficult situations.

Finally, in our summary of motivation highlights, back to education. There seems to be a resistance on the part of some educators toward applying constitutional principle and attending ethical parameters to real world problems.

Too many sophisticated academicians and, for that matter, government politicians as well, consider the handling of socio-economic and technological issues in today's economic climate as unsophisticated or naive when one presumes to platform his thesis on the theological foundation of our U.S. Constitution.

Hence, we must bring the body-intellectual and the body-politic to the viewpoint on religion and science expressed by those more inspired educators. One of our most distinguished educators and his colleagues expressed it particularly well many decades ago, as quoted in the prologue. In essence, they inferred the need to build a bridge between "the pillar of religion" and "the pillar of science" in the context of "world loyalty".

*   *   *

Part II
PURPOSE

ECONOMIC GROWTH WITH HUMANE ECOSYSTEM BALANCE

"I went to the Artisans, for I was concious they did know many things of which I, Socrates, was ignorant. But I observed that even the good Artisans fell into error. Because they were good workmen, they thought they knew all sorts of high matters.*...and this defect in them over-shadowed their wisdom..."

Socrates

CHAPTER FOUR

Reconciling Micro-Thinker and Macro-Thinker

We can now approach the modus operandi of the players bearing on our central humane purpose. Keep in mind the foregoing summations of issues standing as obstacles to **ecosystem balance**, the underlying **principles** involved in corrective action, and, finally, the need to shift the motivations of the decision-making professionals toward the parameters of **humane purpose.** In all of this, we shall attempt to appreciate the not incompatible aspect of a humane ecosystem balanced with profitable economic growth.

In strengthening this rather ambitious approach to the advocacy of a balanced global ecosystem, we must enlarge on certain behavioral considerations as they bear on realizing other than a superficial response to the demands of humane purpose. While all professions are involved to varying degrees, the emphasis here is on the technological, industrial and educational domains. The need is for a massive shift in the response attitudes of scientists, technologists, industrialists and other professionals toward (1) the imbalance between micro-thinking and macro-thinking, toward (2) the lack of an operational perspective of time compression in real world time and toward (3) the spectre of raw materials, commodities and energy process dynamics in the new century.

At the outset, it seems appropriate to once again turn to the pre-Renaissance admonition of that great heretical English philosopher, Roger Bacon. His admonition holds true today more than ever: "Cease to be ruled by outworn dogmas and authorities", he shouted. "Look at the world!"

The grossly overbearing thrust of science and technology toward the microcosm of things has been in vogue just about long enough to warrant the appellation, "Outworn dogmas and authorities". Ceasing to be ruled by them, however—indeed, persuading the micro-thinkers to "Look at the world"—has not been found an easy thing.

We earlier noted the deep plunge into the scientific microcosm during World War II and the impact of that landmark success on the subsequent intensification of the pursuit of the microcosm. Much sooner than normally paced research, this brought in micro and sub-micro technology and, more recently, emphasis on the new nanotechnology. We also acknowledged the good and predestined things arising out of the microcosm of science and technology as we enter the new century with a population of six-billion people growing to as high as ten-billion people over the next half-century.

But at the core of things, a caveat is underscored. In our headlong pursuit of the microcosm, we have all but lost sight of the portent of the enormous scale of the twentieth century's growth and its grossly deficit ecosystem.

The delineation of our approach to humane purpose, then, has its seat in the fact that the "micro" of physical science is in critical imbalance with the "macro" of operational systems integration. So, in our approach to economic growth with humane management of the global ecosystem, we must first address certain of the now deeply entrenched barriers faced by the advocates of macro-system thinking.

Increasingly acknowledged among the contemporary thinkers is an ever-widening chasm between the constitutionally oriented investigators of ethics, morality and Divine Causality on one hand…and the educators, politicians and industrial promulgators of economic growth on the other. Ethics and morality, in the total scheme of things, have drifted somewhat from immediate view. This has yielded a perhaps unintentional but massive divergence of thought in the steering of global enterprise to the ends of ethically founded humane purpose.

While much of this problem has its seat in shifting social mores, there is a missing link in the no longer valid—if ever it was valid—basic education methodology. While much could be said here, the pivotal issue is readily covered. In a few words, it is the teaching of micro-oriented **specialism** at the expense of macro-oriented **generalism** with insufficient regard for interdisciplinary method. While many promising dialogues have taken place in curricular seminars, many of which the writer participated in, not much has happened to strengthen, meaningfully, the interdisciplinary generalist's posture.

Notwithstanding encouraging inroads made into holistic advocacies, the true generalist probes a much wider range of parameters. The generalist is not only a well rounded interdisciplinarian, but is a heavily backgrounded technologist with extensive in-the-field exposure to operational systems. In short, he is super sensitive to sociological and technological macro-system values. Eventually, with an interdisciplinary team facilitated by the OMEGA matrix—or some system closely akin to it—he will orient large scale system and system integration problems to elicit proper response to constitutional principle, humane purpose and ethical operational system goals.

Parenthetically, however, we must quickly acknowledge those highly qualified generalists who have done and are presently doing very productive interdisciplinary work. This is the principal reason for the many operational systems now serving our daily needs. But their tasks are extraordinarily difficult since they are forced to operate without an all-embracing OMEGA-like forum and without, as yet, proper support through generalist training in our education system…The evidence of this is all too glaring in the light of the gross system inefficiencies which have triggered our rapidly deteriorating global ecosystem.

And here we find the crux of things. For the most part faculty and students alike continue largely as victims of the Cartesian paradigm, steeped in the microcosm with a blurred vision of the macrocosm. And for all of the unquestionable accomplishments found in the specialist regime, this blurred image of the macro-realm applies to most of the population of the educational and professional communities.

This is most vividly apparent in science and technology, comprised mostly of an enormous community of micro-thinking specialists. Compounded by layer upon layer of sub-specialists, they pursue phenomena, invention and innovation with very limited understanding of the ethical issues often triggered by their actions.

Moreover, the specialist, aware of his often substantial competency in the micro-realm, too often assumes **unwarranted**_authority to make macro-realm decisions on issues warranting higher level interdisciplinary purview by the macro-thinking generalist. But right or wrong, the micro-thinker often wins his case on the strength of his clout in "getting out the vote" of support from the massive micro-thinking specialist majority over the macro-thinking generalist minority. The alarming subtlety here is that a micro-oriented decision which appears acceptable at a lower level can in too many cases be found wrong at a higher system lavel…We shall later discuss examples of this dichotomy.

Though the specialism of the micro-thinker community is not easy to deal with, the psychology of it all is not difficult to explain. Micro-specialism, for most aspiring professionals, has great appeal since it is emotionally comfortable and offers high comradeship and professional community. Indeed, the specialists will often shy away from opportunities to become generalists. The specialist's lot is usually found very satisfying and often rewarding in the opportunity to know virtually everything about one subject.

This is in sharp contrast to the generalist's often lonely position knowing less about a given subject, but much about many subjects—knowing enough to find the whole greater than the sum of the parts and to challenge the specialist as necessary to assure ethical relevance in the resolution of broad interactive situations. Reconciling this dichotomy of specialism versus generalism is paramount to attaining a high order community posture for the generalist.

This schism between the largely micro-thinker-oriented perceptual present and the macro-thinker-oriented conceptual spread in time is seen with deeper understanding in philosopher Henri Bergson's writings.

Bergson had little patience with those micro-thinkers who advocate stifling logical procedures which deny the flux of experience as it endures in flowing time. The logical procedures of the micro-thinker have their place, of course, but in dealing with the flux of time and experience, macro-oriented operational statements are more to the point. In short, designing correctly for the perceptual present requires first reckoning with the conceptual time frame.

A strong advocate of duration as the substance of form, Bergson was the first to offer the new conceptualism, a way to delimit intuition in the scientific realm by fusing scientific objectivity with artistic directness—micro to macro. Bergson placed considerable emphasis on the importance of looking to those macro-thinking actions responsive to the conceptual dictates of duration and intuition.

It is interesting to note that Bergson's great work had been much misread and was viewed by many as having rejected reason. Yet, much of Bergson's reasoning can be viewed as akin to Spinoza's seventeenth century reasoning, through which one may affirm Bergson's twentieth century thinking as a completion of reason rather than a rejection of reason.

At this point, with further Bergsonian views to be offered in the next chapter, let us pause to examine some examples of the schism between the micro and macro levels of thought—between science and technology and the ultimate operational users of their products.

We will review briefly two kinds of global operational system problems: (1) the ever burgeoning and now critical air traffic control system problem and (2) the also critical global agricultural system problem.

The air traffic control system problem has a comparitively narrow spectrum of design team participants centered primarily on operational flight and related science and technology. The tentative design of a new air traffic control system is now an on going process comprised of a team of federal aviation people, airline captains, air traffic control system operators and a large industrial group of chipmeisters—the computer scientists and technologists. As usual, the operational user group—the macro-thinkers—are a relatively small group as contrasted with literally hundreds of micro-thinkers involved in the design, logistics and manufacturing process.

One can only hope that a strong, macro-thinking operational generalist—preferably a well seasoned airline captain who is deeply versed in computerism—will be put in charge. The present design effort is moving along some rather questionable conceptual lines according to a number of operational people.

Without going into design detail, since many changes will no doubt take place before the new system in placed in service, the basic idea behind the new system is that of "free flight" among aircraft. This means a system governed by computers which measure in three dimensions the relative positions of aircraft to one another, combined with certain of the air traffic control methods currently in use.

Let us pause, then, and take a brief look at the airline captain's task as it exists today, a task often involving severe time compression…The airliner flight crew must operate very much in the **now** of time, a factor usually treated too casually by the chipmeisters, sipping coffee in the cozy environs of the design laboratory. The airline captain in flight must constantly ask: "What is taking place in this hour and how is this hour to be operationally accounted for?" Under time compression of certain airport approach conditions, can the given operational tasks be performed safely and efficiently within the tightly allocated minutes available?

Airline and general aviation operators face those questions hourly as the skys become ever more crowded, and as pilots and air traffic control operators endure a kind of pressure not generally known to the public. Air traffic control operators must suffer the aggravation of pitifully outmoded traffic control systems. These outmoded systems, though updating is planned, are unable to reckon efficiently with the time compression associated with extreme variations in aircraft speeds, rates of climb, altitude and air traffic density.

One of the best kept public secrets is the stream of pilot reports on near misses pouring into the Federal Aviation Administration. A still larger secret is the number of near misses occurring within the overcast and not seen by anyone. The fewness of reported mid air collisions is a combination of the size of the sky, pilots and air traffic control operators doing their vigilant best under less than ideal conditions, and, don't laugh, guardian angels hard at work.

So let us look in on the airline captain and his flight crew as they address their allotted and inflexible operational time to, say, an approach to a heavily trafficed major airport at night in bad weather with an inflight mechanical emergency. Captain and crew have their hands full. Under the proposed new system, they will be required to monitor a new, not easily interpreted air traffic control screen installed in an already crowded cockpit with its nearly one-thousand levers, switches, knobs, instruments, status lights, navigation aids, communication devices, etc. At times, the pilot borders on serious overload in a normal capabilities sense...It is not always a sun filled day...Put the not infrequent emergency situation into the equation and the pilot will, at times, be uncomfortably dependent on the limits of super intensive flight traihing. Airline management simply has no other choice at present but to provide super intensive pilot training to fill the gap between less than optimally designed operational systems and the limits of human capability.

Obviously, the air traffic control system problem with its ground based and aircraft based operational systems and related operating procedures must be seriously addressed. And the problem is being seriously addressed. But of concern is that which has been discussed throughout this chapter—the micro-thinker versus the macro-thinker in operational system problem resolution.

Thus, the key macro-thinker in this problem, the highly seasoned airline pilot, should be the generalist in charge. He should also have an in depth knowledge of computerism in both design and operational terms. He should be well enough versed in the design intracacies involved to reject any system which is too ponderous to be digested efficiently by the flight crew in operational flight, given the relentless time compression factor in the emergency situation noted above.

As has been frequently the case with the products of micro-technology, they do not always perform as insisted upon by the micro-specialist designer—and literally hundreds of micro-thinkers are involved here. The new system, whatever it may turn out to be, must perform with no reservations. Moreover a frail operational system acceptance test must not be tolerated—that is, a test designed by the system designers in which acceptance conditions are idealized to a degree which does not reflect the real world of air traffic control.

A strong minded airline captain generalist must be prepared, authoritatively, to kick chipmeister butt and not be persuaded to accept, however unintentional, half truths by the chipmeisters. Our new air traffic control system, whatever it may finally turn out to be, must perform one-hundred percent—unequivocally!

And now we shall address an entirely different kind of operational system problem—the global agricultural situation. While the agricultural problem, unlike the air traffic control problem, does not hold certain life or death issues in an immediate time frame, it does involve large scale life or death issues on a protracted time line. The global agricultural problem also involves an entirely different kind of operational design and implementation team. Where the air traffic control system problem is centralized to relatively few participants, the global agricultural team, obviously, will involve a very large diverse spectrum of geopolitical, sociological, scientific and technological skills. Nonetheless, however large the team, if great care is not taken, it will be directly or indirectly governed by both the whims and realities of micro-thinking practitioners of agricultural bioengineering.

But first it must be noted that with the incredible diversity of religious and political structures on the global problem level, earlier discussed theologically premised constitutional principles can not always obtain in an approach to the problem. On the other hand, a global agricultural team, say, under the aegis of the United Nations, can certainly be structured to deal with the inescapable bottom line—global survival and increasing hunger.

Now, team aegis identified, let us examine the unique role of the yet to be appointed generalist team leader. First, after appropriate consultation with key participants, he must appraise the kind of world we face and build consensus toward the global agricultural problem.

Published views on the general character of the future vary widely. Some commentators see a world with a widening chasm between the haves and have nots as predatory capitalism further penetrates the world milieu in search of cheap labor without taking responsibility for the ambience in which that labor exists. Other commentators see a world expanding from the status quo to even greater economic glory. Both views reveal a surprising lack of regard for further billions of people yet to be added to an already hungry world. Then, there is a third view, a penetrating but not yet well supported view of world hunger. This is seen as a worsening human condition through billions of mouths to feed as many global rivers dry up, as water tables fall on all continents and as additional agricultural land placed in production falls far below present as well as future population needs. World hunger will soon be on a galloping increase.

Enter, then, our not yet appointed world agricultural team's leading generalist, who not only has a firm grasp on the particulars of world hunger, but who has the statesman-like ability to approach the political and socioligical obstacles to dealing with the agricultural problem on a grand scale and to build consensus for proper support for what must be done.

With a background of in depth knowledge of the research and development capabilities of the bioengineering microscientists and technologists working on the agricultural problem, he takes his first step. The generalist rallies the pockets of science and technology throughout the world who have addressed themselves to agricultural genetics—the micro-specialists who are making significant inroads into genetically engineered seeds, plants and agricultural practices.

We will not at this point discuss the enormous team structure involved in reckoning with global agriculture, but rather will focus briefly on the highly commendable efforts of the genetic scientists doing this fine work. To mention only one example: Corn and corn byproducts bioengineered to improve quality and to protect from pest infestation, using genetic information as a substitute for insecticides and pesticides. Indeed, a two-for-one approach which will attenuate pesticide pollution while increasing food production at the same time…Encouragingly, there are many other productive bioengineering efforts now in process to improve global agriculture—faster growing more productive plants, demanding less water and offering greater adaptability to less than optimum quality agricultural land, and so on.

The generalist team leader, however, has the awesome task of not only building a supporting consensus across a wide spectrum of political and sociological institutions, but he must build a bridge between what the micro-scientists claim for their products' performance in a laboratory environment and the exact conditions of the often harsh, specific agricultural environment in which these products are to perform. The generalist must also ascertain, in very real terms, the production volume and logistics potential of laboratory crafted agricultural products…In short, the generalist team leader must remain most alert not to accept any pie-in-the-sky claims by the micro-scientists—claims which at the laboratory level are justified, but which at the volume production and actual field environment level may not be justified.

Another of the principal tasks of the team leader will be to differentiate between those macro-values which are wholly technological and those macro-values which are sociological or political; and which, in turn, will often influence the approach to implementation of new methods.

Finally, the team leader will maintain an intense alertness to the state-of-the-art in bioengineered agriculture and seek further opportunities to move the hunger pangs of the world beyond a few islands of privilege.

In closing this chapter, it must be underscored that in strengthening the macro-thinker's posture, we must deal with the hue and cry from the micro-thinkers to the effect that they are aware of the macro-system. But this, in most cases, is merely cerebral, not from the heart. In all areas of science and technology, the micro-thinker, on one level or another, has almost always ruled the research, development and operational implementation situation—and often with far from optimum results at the operational user end of things.

It is appreciated that our discussion of the micro versus the macro has been somewhat vigorous in urging a massive shift in the thinking of the micro-oriented professionals toward an attitude of greater responsiveness to the massive macro-system dilemmas facing the world today. This means greater responsiveness early on to the contemplated whole of a massive operational system…and not quite as much force fitting of macro-thinking after the operational fact.

We do indeed need the micro-thinkers and the marvels they bring to the fore. But let the practices of science and technology and the educational processes of academia come together to reconcile the micro-thinker and the macro-thinker to an ethos of humane purpose.

*   *   *

CHAPTER FIVE

Reckoning With Time and Time Compression

Time and time compression considerations have been highlighted throughout the text thus far. It is now appropriate to reckon in greater depth with this vital, all important, yet most common dimension of life and growth. Indeed, the homely clock is perhaps the most enduring, unchanging man made devise for control we have ever known.

Remaining briefly in this fundamental context, a composite of various dictionary definitions of "time" will serve as a discussion platform: Time is **duration**, continuance, in which things are considered as happening in the past, present and future...Time is every moment there ever has been or ever will be, **a system of measuring duration**...Time is a period characterized by prevailing condition or specific experience...Time is the allotted period during which some act is to be performed...Time is that point at which something has happened or not happened...**Time is a medium for adjusting or setting events so as to coincide in system concert with one another to serve some specific objective.**

Obviously, the above definition speaks only to the real world of natural time rhythms. That other definition of time—the often "unreal" or "surreal" nanosecond world of computerism—and the caveats implicit in it, will be enlarged upon after some further comments on Philosopher Henri Bergson's views of time and duration. Bergson has set forth one of the most lucid and enduring treatments of the phenomenon of time and man's actions within it.

He speaks pointedly of the methodology of science and its propensity for arresting time so that it may clinically examine the things it perceives...but without entertaining the bothersome dynamic associated with the fact that the thing in question normally moves and endures in the reality of some whole environment.

Bergson gives us a platform on which to speak of this often incorrectly viewed divisibility of things. Now prevalent in the microcosm of technology is the incorrect assumption that things and situations can be arrested in time—removed from the flux of experience, lifted out of duration in flowing time and acted upon in isolation. In Bergson's words: "Science eliminates duration from time and mobility from motion before it can deal with them." Now, in itself, this is not a bad thing. It is a most necessary scientific technique—that of isolating some element of a larger system so that it may be statically examined in the microcosm. Too often, however, a preoccupation with the microcosm sets in and the element in question does not become legitimately subject to reidentification with its parent dynamic in the mainstream of time.

It is readily seen that the scientific microcosm has become a bedfellow of that natural propensity of all men, to allow themselves to become swallowed up in the perceptual attractions which dominate the now of time. That things and situations can be held static and divisible—perceivable on a day to day basis—causes a leaning toward the practical expedients in this. In limiting one's view to a day-by-day basis, things become immediate, subject to task allocations implicit in commerce. In this context, things are taken apart and shared as in the breaking of bread.

Consider, for example, the design and implementation of any large operational system by the prime contractor for the whole system. Such programs often involve subcontracting to many scientific, technological and industrial firms. While the prime contractor for the total system parcels out the specifications for the specific tasks he is subcontracting,

those tasks are often specified with less than full exposure to the subcontractor of the full scope of the total program. Thus, lack of full exposure combined with the subcontractor's own motivation means that the
end point operational performance objectives of the system as a whole
are not always realistically approached.

Moreover, the subcontractor's view of his own profit oriented piece
of the total program is so circumspect that, for him, real continuous
time duration in an operational system user context does not even exist.
He simply addresses himself to profitable containment of a narrowly
circumscribed part of some larger whole.

The result? All too often the subcontractor's time arrested part is not
found optimally compatible with the larger system it was intended to
serve. His part only combines with other subcontracted parts to form a
patchquilt of machine clusters and the user's operational people. This
means that the system as a whole is operational to a degree, but with a
measure of inefficiency which often fosters serious entropic deficiencies.

The foregoing brief discussion of Bergsonian time philosophy
applies on many levels. The point to be stressed at this juncture is a
point related to humane purpose. Regardless of the micro-level with
which technology may be concerning itself, it is the role that micro-
effort is destined to play in some higher order macro-scheme that is of
concern—a macro-design not always drafted in such a way as to control
the micro-effort. In short, the micro-view of things ought to be ruled by
the perspective of the ultimate operational user of the system.

Let us now examine "time" more closely in the context,of time compression. Consider the degree to which our explosive economic growth
situation has suffered a lack of response to the emergence of time compression. Note that as the decades of the second half of the twentieth century passed, a severe pile up of incorrectly handled operational events
occurred across a number of various systems. These incidents resulted in a
severe increase in the rate of system malfunctions, failures and, most disconcerting of all, human error. This, not so simply, is a matter of system

designers failing to provide **operationally coherent designs** which would not overload the human operators in a climate of ever increasing loads on the system.

If sufficient clock time is not available for dealing with a given set of operational actions, the human operators involved obviously have but one choice: Learn to handle those actions in less time—in effect, compress time. This, as the record clearly shows, is not consistently possible.

Now, some readers will tend to challenge the emphasis placed on time compression and the general frailty of our operational systems. But there are two sides to the time compression syndrome. We have on one side of the coin addressed time compression largely in the context of specific now-time tasks by the people who operate the systems. We must now discuss the equally vital other side of the time compression coin—the operational systems themselves and their accompanying operating status.

Putting it too simply, perhaps, there is a monsterous subtlety at work in general public and professional attitudes toward our operational systems at large. Our systems appear to function—albeit very inefficiently—as they crudely serve their daily purposes...and, for the most part, continue to serve their purposes. After all, constant improvements are in process and we seem to be muddling through...What, then, is the great concern all about?

The general public is the last to know of the magnitude of this problem. Yes, the news media regularly broadcast the latest oil refinery and dramatic fire accidents, the electric power system failure bringing one of our largest cities to s standstill for nearly a day...caused by a combination of human error and a patchquilt power distribution system now undergoing a major redesign, a rotted major water system breakage causing severe damage, a major rupture in a natural gas line, the railroad disaster caused by faulty switching, the near meltdown of a nuclear power plant...and so forth.

Now, when we say the public is unaware of the magnitude of the problem, it is because there are literally **thousands** of accidents and operational system failures every year that do not reach the level of "newsworthy", but which, cumulatively, find our general operational systems suffering severe time compression and contributing heavily to our deteriorating ecosystem.

What all this means is that the entire world is suffering critical "entropy leakage" from its myriad operational systems. We are daily falling further behind in reckoning with this leakage. All systems, of course, serve us with some degree of entropy...but total entropy from all systems is no longer acceptable. Fixing the whole global enterprise requires much more effort than has been allocated. We suffer time compression, yet it is offered that much more effort is spent perpetuating an intolerable situation than addressing the problem in depth...and then acting in a more timely manner.

And from the standpoint of initiating timely action, we must make note of the degree to which top politicians are compromising ecosystem issues to the ends of political expediency. The Vice President of he United States,in a 1998 television address to millions of people,was boasting about how well our ecological posture is coming along. As one illustration, he spoke of the great efficiency of the new skyscraper buildings brought about by the environmental movement. These new buildings were not only touted as more energy efficient, but he made a special point of how much lighter the new buildings were because of design advances which required fewer materials and thus sharply reduced demands on our natural resources base...Wrong! While what he said is true of the new building per se, he failed to mention the net **increase** in resource use as the old, inefficient skyscraper was torn down to make way for the new skyscraper. Consider the logistics and resources drain on **the project as a whole.**

This involved demolition equipment, hundreds of truckloads of heavy debris and, cumulatively, thousands of truck miles to an often distant disposal site, extensive grading operations at the disposal site and thousands of gallons of fuel. All this just to rid the construction site of the old building. Moreover, he failed to mention the logistics of the heavy equipment, materials, trucks and fuel involved in the new construction, let alone the energy and raw materials resource chain behind the production of trucks, construction equipment and fuel production...not to mention the substantial entropy. **Legitimate statistics** require proration of all logistic, raw materials, energy and entropy factors applying to any given project.

Innocent or not of untruthful statistics, politicians and other influential public speakers are approaching the general public with statements about good things happening which all too often are but grossly misleading half-truths.

The net result of all of this is the unmonitored further deterioration of the ecosystem. Some of the entropy, of course, is unavoidable. But without question, awareness of the ecosystem issue can lead to meaningful entropy reductions.

We must pause here for a parenthetical on this matter of "time" and summarize briefly the three categorical modes of real world time. These are cosmological time, conceptual time and perceptual time. Discussion following this will contrast real world time with the nanosecond world of the microcosm and the roles in time compression played by both real world time and microcosm time. First then:

Cosmological time, for OMEGA purposes, we quickly dismiss as remote from the immediacy of our thesis. That is, we are not concerned here with earth's ultimate eschatology—the cosmological frame of reference wherein the earth is expected to cool and die some billions of years hence. We are concerned here only with that cosmic speck of time embracing the great curve of population growth and its immediately related conceptual time and perceptual time perspectives.

Conceptual time perspective addresses problems in the context of that which is involved in the operational interlock of past, present and future…and asks the question: What is the impact of time compression on that span of calendar time being investigated? For example, consider the problem of waste disposal wherein the conceptual time frame is placed to either side of the present. In short, technology thoughtlessly buried its waste in **the past.** That waste is leaching its way into **the present.** And in a climate of severe time compression, waste disposal is not being dealt with vigorously enough to safeguard **the future…**More later on a few good accomplishments in waste management versus the gargantuan task yet ahead.

Perceptual time perspective, on the other hand, is that mode of time which deals with **the now of things…**and asks the question: What is expected to take place in this hour and how is this hour to be operationally accounted for? Under time compression, can the given operational task and its often many sub-tasks be performed within the sixty minutes contained in that hour? Those questions can be posed with varying degrees of urgency across the wide spectrum of every operational system serving our daily needs.

Most of the vast array of operational systems serving the global enterprise suffer to varying degrees from task overload or excessive entropy. This underscores the importance of full conceptualization of an operational system before approaching the design of micro-oriented sub-systems and components. This is underscored because of the wide practice of attempts to save time by getting into component design before system conceptualization has been sufficiently mapped out. This practice is acceptable only to the degree that it is more carefully monitored for possible negative downstream effects than is presently the case. All too often, time is lost rather than saved…and entropy continues unabated…More later on the hazards of operating nuclear energy systems and various other utilities and industrial systems not currently geared to the ever increasing tempo of life and attending time compression and entropy constraints.

The pivotal answer to all of this, though not-so-simply, lies in sifting relevant information and executive actions from the present crushing burden of incoherent nonrelevancies.

As previously noted, the reader will find this book a strong advocacy for the macro-world of the Clock Culture, of the natural rhythms of real time as we daily live it. It is not an advocacy for the micro-world of Computopia, the nanosecond life of unrestrained information flow in which response to relevant human need is often only happenstance, and in which change for change's sake becomes the new ethic.

The keynote of this book is the seeking of relevance to the conditions discussed throughout the text, drawing upon the, yes, **relevant** wonders of micro-science without being ruled by a **nonrelevant** micro-ethic which allows micro-creative processes to roam freely and find their own order.

Now, since the whole idea of a new initiative for differentiating relevant and nonrelevant issues is premised on reconciliation of the advocates of the Clock Culture and the advocates of Computopia, recognizing the sweeping political and industrial dimensions of this, it is appropriate to briefly note the differences in points of view between the two.

The Clock Culture has its seat in the natural time rhythms of nature, when it gives pause to reckon with this, and is the appropriate steward of our global ecosystem posture.

Computopia, on the other hand, places its emphasis on subsuming the natural time rhythms of the Clock Culture…and even worse, as previously noted, Computopia is moving toward a micro-ethical behavior of its own.

At this time, as we enter the new millennium, the foothold that Computopia and micro-science have already gained to that end is awesome…and alarming. Daily, it becomes more difficult to strike a balance between the Clock Culture and Computopia schools of thought.

In brief summary of this situation, it is, yes, the many good things given to society by the micro-scientists of Computopia which have, thus far, unwittingly steered the administrations of government and industry away from the magnitude of the real problem—that of overcoming our rapidly deteriorating global ecosystem. Computopia's contributions to handling time compression, while limited in the scale of the global problem, are nonetheless significant in medicine, accounts processing, machine control systems and so forth. And while insufficiently supported thus far, it is acknowledged that there are other productive micro-science efforts in energy source research, agriculture and other fields addressing problems which can lead to improvements in the ecosystem.

Yet, even in that context of good accomplishments, there is growing conflict over how best to handle certain situations. The point is, Computopia's and micro-science's good accomplishments have accorded them an unwarranted political strength. It is offered that the true magnitude of the time compression and ecosystem problem can be realistically approached only by the macro-thinkers of the Clock Culture. They possess an inherent affinity for the natural rhythms of time and the macro-ecosystem problems that attend the real world.

What is needed, then, is a highest level policy agreement which, in the name of a survival oriented ecosystem, would put Computopia and micro-science clearly under the control of the macro-thinkers of the Clock Culture. The purpose of this, of course, would be to serve high order ethical prerogatives which would logically subsume the micro-sciences of Computopia...instead of the present opposite trend wherein micro-science reaches for greater control than that which it has already assumed. (Jeremy Rifkin in his excellent book, "Time Wars", suggests "time" as one of "the primary conflicts in human history". Very important reading.)

With all of the foregoing in mind, then, we must move toward a philosophy and method through which both of these adversarial factions can reckon with parts, whole systems and actions as elements of the Bergsonian "indivisible main stream of time"—real world time—flowing through past, present and future. Then, too, only with the indivisible mainstream of time in mind can we properly embrace a workable view of time compression. Complex and unprecedented group actions are involved here, of course. Present concepts of task allocation, responsibility and the ultimate award of profit will shift markedly as we place emphasis on continuity and integration of operational systems and, in turn, ecosystem refinements aimed toward achieving true ecosystem balance.

Conceptual actions then, representing an interlock of the people, things and events of the past, present and future need not be approached through sequential reactions to the perceptions of the moment—as is too often the case. These perceptions of the moment are often offered as conceptual actions and sometimes do appear on the surface as conceptual actions, but the march of events in the tempo of our era in continuous time flow soon reveals them as frail ad hocisms.

To implement the kind of conceptual actions needed in our era, the sharing of the pie of economic growth must occur on a macro-system basis in a climate of understanding that the whole of the environment of things—now flowing in ever compressing time—must change the very character of those things; and at a rate which makes advance OMEGA-like planning and control mandatory.

Parenthetically, we should not pass this point without acknowledging technology's promising inroad into scientific reckoning with the dynamics of conceptual time. This promising inroad is being effected by the science known as operations research. This science is grappling with some of the aspects of the dynamics of continuous time flow with most promising results. Not only do subsequent chapters of our discussion draw on certain of the techniques of operations research, but, hopefully, even add to its methodology as it bears on addressing enduring problems of great magnitude.

However, notwithstanding technology's heartening trend toward dealing with the dynamics of real world time through the science of operations research, the overwhelming propensity of technology at large is still aimed toward the time-arrested microcosm of things or to the profit structures of the perceptual present. In short, for all of its potential, operations research does not yet have sufficient voting power. In this regard, the long tenured term, "operations research", falls short semantically of the all embracing dynamic of the term "operational time flow". Moreover, most macro-system research functions addressed to conceptual time flow problems are severely constrained by lack of supporting policy.

Hence, our humane purpose seeks, among its several high order goals, a platform for inversion of emphasis in the science and technology community…placing in the immediate foreground an ethical view of operational time flow in the context of ecosystem balance. This means nanosecond Computopia and micro-science in, yes, vital, but subservient roles.

The parts and subsystems of the operational ecosystem as a whole— both the natural and man-made elements comprising the whole— must be seen as elements in a concert of motion in the relentlessness of continuous time flow. We must respond to Bergson's earlier admonition and corrective philosophy, urging science to **"put duration back into time and motion back into mobility."** if humane purpose is to be served properly.

*   *   *

CHAPTER SIX

Responding to 20th Century
Growth Dynamics

It is assumed our discussion thus far has reasonably enhanced an appreciation of the macro-view of the indivisible mainstream of time as well as the other dimensions of the global ecosystem bearing on humane purpose.

We can now proceed to examine the dynamics of the whole twentieth century and its unique phenomena of growth, evidencing growth characteristics unlike any other period in history. This chapter will examine humane purpose issues in the context of the last half of the twentieth century as they are likely to relate to the twenty-first century. We shall examine the twentieth century's (1) population growth versus (2) raw materials/commodities demands versus (3) energy conversion demands as shown on Figure Two. Rather than attempt a quantitative whole world statement at this point, it is offered that the U.S.A. data comprising the three curves can be qualitatively extrapolated in principle to the whole industrial world and the developing nations.

A few words on the origin of the smoothed-curve totals shown on Figure Two are in order. Those curves and accompanying observations cost over a million dollars of corporate general research funds, involving an interdisciplinary team of ten people over several months. It was the writer's privilege to direct that study. At the time of the research many years ago, those curves revealed a virtually ignored economic growth behavior phenomenon—a phenomenon which even today is not, to the writer's knowledge, adequately embraced by academia nor by the professional community at large.

TWENTIETH CENTURY ECONOMIC GROWTH DYNAMICS
(These curves are based on U.S.A. data
but in principle apply also to the world situation)

1. POPULATION GROWTH-(Including demograpbic factorial)
2. RAW MATERIALS/COMMODITIES DEMANDS
3. ENERGY GROWTH DEMANDS

These deceptively simple appearing semi-qualitative exponential curves are comprised of many hundreds of data fragments converted to percentages and smoothed-curve totals. They represent the three highest level operational elements in economic behavior—population, raw materials/commodities, energy—an economic behavior statement to which all economic theories and practices must be subordinated. Once these heretofore unexploited dynamics of twentieth century growth are formally embraced, and understood in the context of time compression we can begin to move toward a structured OMEGA matrix approach to economic and ecosystem equilibrium on a global scale.

A carte blanche approach to this study was authorized by top management with a view toward stepping outside conventional economic growth analysis to see if we could ascertain heretofore undiscovered broad based business opportunities that lay on the distant horizon. It may be of interest to note that these curves are now over three decades old. Yet, the passage of time has only confirmed these curves which, at the time, were to a large extent speculative projections from very spotty data spread unevenly across the whole century and many industries. Not only are these curves qualitatively valid today, but will come to be found conservative. This will become evident in reckoning fully with the entropic dictates of pollution containment as we face into the largest "commodity" of all time—waste. Our original projections simply did not account for the true magnitude of the waste and pollution problem…nor the severity of other growth factors to be discussed later.

Now, the uniqueness of these deceptively simple looking curves suggests that a brief review of the original construction of the chart might be of interest, particularly from a standpoint of our concurrent search for ways and means to present complex issues in the clearest possible terms to bright but uninitiated top management decision makers.

We first elected to view many hundreds of data fragments through a common window—the whole twentieth century. With the horizontal dimension of the large six foot by ten foot chart struck off in decades, we then posted the backbone of chart. We posted the U.S.A. population growth in millions of people including all U.S. citizens working abroad in the more than one-hundred foreign countries from whom the U.S.A. acquired raw materials or exchanged goods and services. The population curve also reflects a demographic factorial, qualitatively integrating absolute population growth and the massive migration of people from rural areas to urban areas after the beginning of the twentieth century.

The vertical axis of our large wall chart was a multi-set of scales forced by the initially uncorrelateable hundreds of data fragments with no common standard of comparative measure. As a reminder, note just how disparate these vertical scale measuring terms for raw materials, commcdities and energy demands really are. Here are a few:

Kilowatts of electricity. Barrels of oil. Cubic feet of gas. Tons of coal. Acre feet of water. Board feet of lumber. Cubic yards of cement. Tons of steel. Bushels of grain. Bales of hay. Gallons of fuel. Quarts of liquid. Yards of fabric. Acres of land. Tons of earth (moved or graded). Etc., etc.

Nonetheless, with the big wall chart and its whole century baseline and population growth curve, we set out to collect and post whole century actuals and projections. In the course of this study, we called on literally dozens of agencies in both the government sector and the private sector, and to the best of the writer's knowledge, no single agency is even today compiling **fully integrated** comprehensive statements of the kind reflected on Figure Two.

In any case, with nonstandard measuring indices bothersome enough, each measurement in its own right proved fragmentary—a ten year increment here, a fifty year increment there, with much historical data not reaching back to the year 1900. Then, too, much of the projected data were either almost nonexistent or sporadically treated in five, ten, twenty year increments and so forth. Only a few fragmented projections other than the population growth curve reached the end of the century at the time of the research.

The question some will ask: Was not the collective data virtually meaningless with their nonstandard measurement indices and fragmented time scales?...Yes, but for our sensitivity to serendipitous and intuitional input. A few months into the study, the huge wall chart seemed an amorphous growth of undecipherable Mumbo Jumbo. But! The large room in which the chart was posted permitted a long distance view. Standing back twenty feet, too far to read off a particular data element, a phenomenon revealed itself. With the extra heavy black line of population growth running across the chart on which the whole century was posted, one could see hundreds of data fragments on raw materials, commodities and energy growing essentially in parallel with the population curve during the first half of the century. In the second half of the century, however, that amorphous collection of hundreds of data fragments presented a blurry image of growth crossing above the population curve.

The next step was by then an obvious one, albeit somewhat arbitrary. We changed the vertical axis of the chart with its many measurement indices to percentages and elected mid-century as the 100% point not only for population growth but for all data fragments. This massive posting at once eliminated uncorrelateable modes of measurement and brightly illuminated the explosive growth crossover of raw materials/commodities demands and energy conversion demands after mid century.

We then took a further step toward enhanced clarity in order to present that portion of our research in the most basic terms. We gathered all raw materials/commodities demands into a smoothed average, single qualitative curve and then did the same for all energy conversion demands.

Our now decades-old chart stands firm with earlier projections essentially validated except for the previously noted largest "commodity"—waste…which threatens to further steepen the curves with which we must reckon in working our way out of the energy versus ecosystem climacteric.

While we shall discuss further the various aspects of the explosive crossover and its time compression corolaries, a brief explanation of the underlying meaning of the crossover phenomenon, and the economic growth behavior inferred therein is offered here.

On a per capita basis, a unit of food, a unit of housing, or a unit of utilities must be viewed three ways. These commodity units must not only be viewed in the context of population growth, but in the context of the more rapidly rising raw materials and energy growth demands associated with producing those commodities. These three frames of reference—severely exponential population growth, raw materials growth demands and energy growth demands—while deeply interrelated, of course, no longer parallel one another in a one-to-one relationship. That is: Viewing the middle of the twentieth century as an approximate index period, as shown on Figure Two, we can look back toward the beginning of the century and note essentially one unit of population growth demanding one unit of energy to produce one unit of commodity. But not so after mid-century. The units of energy demanded per unit of commodity are increasing much more rapidly than the concomitant unit of population growth. The pre mid-century's approximately one-to-one ratio of growth in population units to growth in raw materials units and energy demand units had shifted explosively.

Hello, Malthus…

Today, a number of units of energy are required to meet the raw materials/commodities demands of only one unit of population. And the multiple units of energy required to reckon with each unit of population growth threatens an increasing ratio as we move more deeply into the twenty—first century with the energy versus ecosystem climacteric strapped to our backs and as an additional four-billion people enter the global scene. Obviously, these ratios will vary between the have and have not nations, but the sharply increasing trend applies in principle to the world as a whole.

The main point here is this: Efficiency improvements and conservation efforts which would reduce the units of raw materials and units of energy demand per unit of population growth are important and have already helped to some degree. However, in our climate of continuing explosive growth, these efforts, as presently defined, will have a pitifully small overall effect on the essential nature of those growth demand curves. Though generally not recognized as such, the dynamics of the raw materials/commodities and energy demand curves, in the ambience of the new millennium, will endure as **natural phenomena,** even though hardly visible until after mid-twentieth century. This chapter will be sufficed in viewing briefly the basic interacting factors involved.

The now rapid increase in billions of people and the ever widening spectrum of simultaneously evolving new technologies combine to demand extraordinary increases in energy output. Not only do we have the energy demands associated with more people demanding more commodities and services, but we have the still larger component of energy demand associated with more people, more commerce and more government moving about the world in randomly changing life styles and modus operandi…And, to mention again, the demographic effects of the massive migrations of rural populations to the urban areas, which have persisted since early in the 20th century. More movement and faster movement of people, raw materials and finished products from one place to another is demanded, triggering myriad new energy needs as high-energy-demand complex systems are invoked to replace simple systems. For example, the enormous raw materials and energy demands of the not so simple freeway overpass complex versus the simple stop sign intersection…Or the frightful energy price being paid to compress time with speed as passenger air miles and fuel-thirsty jet transports continue to expand. And, of course, myriad other time compression effects and the stuff of growth consumed by them.

Too, more people, more commerce and more government in high density environments incite more social abrasion of all kinds—be it crime, poverty or pollution—with attending demands for large increases in public services, systems and equipment. Most of the above translates into exponential increases in the tons of harder to get raw materials moved in behalf of each individual each year…all of which, of course, translates back to the prime mover—energy!

Now we come to the least publically acknowledged and yet the largest single demand for energy—the energy required **just to get more energy!** That is: Before moving a single unit of energy output into the private consumer sector, we grapple with an incredible raw materials/energy demand spiral. We demand more raw materials to get more energy to get more raw materials to get more energy…and so on into a downward spiral of critically diminishing reserves.

Consider that the first use of a unit of energy is to acquire services and raw materials to design, construct and fuel the power generating turbines. The second unit of energy is used to design, construct and operate the energy distribution system. The third unit of energy is used, or should be used, to design, construct and operate waste control systems together with the monitoring of the overall ecosystem condition resulting from energy generation, distribution and waste control. In short, it drains energy to make energy, it drains energy to distribute energy, and it drains energy to control the entropic effects of energy output. Only after all of that can the ultimate consumer turn on his lights and operate his myriad systems. The ultimate consumer nets barely a third of the total energy produced through the present inefficiently wide spectrum of energy systems used to produce that total energy.

For the foreseeable future, the demand for energy will increase exponentially above the ever steepening increase in the population of the industrialized world. This is an enduring reality. The population growth curve, the raw materials/commodities demand curve and the energy demand curve will not change qualitatively in their dynamic behavior, though subject to quantitative variation as efficiency improvements are realized.

The best we can hope for is to generate a movement toward differentiated, or more succinctly, **discriminate** economic growth. By "discriminate" we refer to the need for much greater investment in the systems development and geopolitics of global ecosystem control. This means more **relevant** selectivity before adopting certain scientific and industrial programs—programs which often proceed with change for change's sake without relevant regard for the whole of things.

We must realize that if ecosystem balance is ever to obtain on a global scale, the industrial world cannot sustain the indiscriminate elements making up a large part of the upward slope of those curves as they are presently constituted. While we pointed out that, qualitatively, the dynamic behavior of those curves is here to stay as economic phenomena, the slope of the curves and their **content** can, with time and dedication, be adjusted to a point of global equilibrium.

In any case, the growing complexity of global ecosystem management imposed by such exponential growth phenomena is quite enough load to carry. But when subjected to unstructured indiscriminate change, the dynamics of energy growth are grossly aggravated, fiercely compounding already severe energy demands. Integral to our humane purpose, then, our approach to dealing with the energy versus ecosystem climacteric, and eventually containing the ecosystem countdown, is to inhibit indiscriminate economic growth on one hand while, on the other hand, not letting up on the realities associated with fulfilling relentless raw materials/commodities growth demands and energy growth demands.

This not yet well understood task is, of course, of incredible scope. Yet the goal and basic approach to the task is clear. As the constitutional ethicists and macro-thinking interdisciplinary generalists subsume those micro-thinking scientists, industrialists and politicians who, perhaps even unwittingly, are indifferent to humane purpose, we can then proceed to structure our economic growth demands. That is, we can **relevantly structure** these demands within the parameters of sound, interdisciplinary global ecosystem management.

Eventually, indiscriminate economic growth will begin to wane and quantum increases in energy efficiency and relevant system development will follow...Again: Idealistic? Yes...But any movement toward humane purpose on a grand global scale will be so identified. But if the principles are put forward by informed educators and research organizations with unrelenting persistence, the parameters of humane purpose will gain a foothold...not quickly, but inevitably.

         *   *   *

Part III
PRECEPT

PUTTING ECOLOGICAL RUDDER TO GLOBAL ENTERPRISE

"Make for thyself a definition or description of the thing which is presented to thee, so as to see distinctly what kind of thing it is in its substance, in its nudity, in its complete entirety; and tell thyself its proper name, and the names of the things of which it has been compounded, and into which it will be resolved. For nothing is so productive of elevation of mind as to be able to examine methodically and truly every object which is presented to thee in life and always to look at things so as to see at the same time what kind of universe this is and what kind of use everything performs in it, and what use everything has with reference to the whole, and what with reference to man; what each thing is, and of what it is composed, and how long it is in the nature of this thing to endure which now makes an impression on me, and what virtue I have need of with respect to it, such as truth, fidelity, manliness, simplicity, contentment and the rest…"

Marcus Aurelius
Emperor of Rome 161 A.D.

CHAPTER SEVEN

Rationale For The OMEGA Matrix

As seen from a space ship, the planet earth is a pristine blue jewel, spinning its way through the heavens as it circles the sun. In the cosmological sense, perfect equilibrium obtains.

But as we put the zoom lens on earth's thin skin of life known as the **Biosphere,** we find a situation far from equilibrium. As the occupants of that fragile thin Biosphere, with feverish intensity, randomly probe the **Geosphere** for subsistence and economic growth, that Geosphere not only drifts further from equilibrium, but impacts with high entropy on that collective community of man known as the **Sociosphere.**

The **Geosphere-Biosphere-Sociosphere** axis of thought may properly be considered, in the context of hierarchical order, as the pragmatic ultrastructure of our earth's basic output systems. These foundation stones of the OMEGA matrix will be addressed and added to following further discussion of some of the basic tenets of the interdisciplinary approach to things. Emphasis will be placed on the interdisciplinary method in the context of a rigorously disciplined concept of hierarchical order. But first, adding to our previous discussion of micro-thinking, we must outline a vitally important limiting factor found in the professional life of the micro-thinker today.

Specialists within the same field often have difficulty communicating with one another, let alone finding bridges to communicate with specialists in other fields. Consider the all pervasive field of mathematics, for example. At the opening of the twentieth century, an individual mathematician could embrace in one head essentially all the known mathematica of that period. This is now an impossibility, as the field of mathematics has branched into sub-specialisms vhich often require collaboration with other mathematical specialists to resolve the problem at hand. This is also the case with other specialisms throughout the microcosm.

And we have yet to fully address the difficult role of the decision makers who must take the broader view, often cutting across many disciplines. As previously noted, the often brilliant decision makers at the highest levels—those who are charged with steering the world economy—are rarely equipped with the mental tools which would enable them to challenge the thinking of the chipmeisters and other micro-specialists.

All this makes absolutely imperative the construction of a cohesive interdisciplinary bridge to a hierarchically structured holistic view of things. This would facilitate a highest order, purpose oriented community of interdisciplinary experts to whom the decision makers could turn for counsel. Although it is encouraging to hear increasing dialogue on the importance of interdisciplinary communication, we can observe very little focused action thus far. Interdisciplinary communication, what little there is when compared with the need, and almost all related decision making processes continue far too much at the mercy of the practitioners of higher mathematics and the micro-thinkers. The critical concern is that very few top executives, and most particularly our political and industrial leaders, are mathematicians, let alone conversant with computer manipulation of mathematical models.

Conversely, the practitioners of higher mathematics as well as the micro-specialists, rarely have an adequate grasp of the often wide spectrum of nonmathematical, higher order qualitative issues bearing on the decision making process.

Yet, no hierarchically structured analytical model of any great scope stands ready to facilitate orderly decision making processes by those multi-faceted, often wise, but nonmathematical leaders who steer the direction and quality of economic growth. Worse yet, through lack of a truly interdisciplinary communication platform, many decision makers allow themselves to be swayed or even intimidated by higher mathematics and micro-specialist thinking. This dangerously retrogressive situation—high level decision makers often found under the spell of the chipmeisters—has persisted for many decades.

The OMEGA matrix, in addressing the above situation, is not only structured to reckon with the Niagra of increasing entropy in the global enterprise, but OMEGA also seeks to reckon with a corresponding Niagra of right information and wrong or nonrelevant information. This situation hungers for sorting and integration in some hierarchically structured, synthesized scheme of things. OMEGA would be aimed toward offering the decision makers the clearest possible delineation of the issues under consideration. The foregoing brings the discussion briefly back to our rationale for election of the Geosphere-Biosphere-Sociosphere axis as, under God, the hierarchically highest order element of the global enterprise. Hierarchical order is, indeed, the governing design principle in our rationale for the Omni Matrix, Economic Growth Analysis…OMEGA.

Rather sadly, though, it seems there has been far too little discussion of hierarchical order since Plato spelled it out in Plato's Hierarchy of Good over two millenniums ago. Plato's hierarchism made clear that at the top of any hierarchical order, all other things which man might consider should be ranked in order of importance. Over simply, any item or issue in a hierarchy is positioned in terms of whether it should govern or be subordinate to the other.

Hierarchical order—one of the most vital of all principles in any planning or analytical process—is barely scratched in today's education system and in the practices of the professional specialists spawned by the present education process. Consider a recent stark illustration of this in dealing with high order analyses of our global future.

A large world oriented organization very much respected by this writer—The World Future Society—offered as one of its 1999 major conferences a program which strayed far away from a desperate need for hierarchical order. They failed to present a hierarchically structured **teachable**, intellectually **trackable** and issue **actionable** program to a point of focused and properly prioritized recommendations.

Quoting their original announcement of the conference program, they offered that: "Issues have been clustered into several broad spheres: "The Biosphere", "The Technosphere", "The Sociosphere", "The Econosphere", "The Politisphere", "The Futuresphere", "The Unisphere".

Now, from a standpoint of lack of hierarchical order, here is the scary part. They offered that "The boundaries of each sphere are fluid; since future issues are by nature interdisciplinary, it is impossible and unproductive to tightly categorize them. Within each sphere", they suggested, "are the issue areas, such as health, education, work and careers, information technology and governance".

Of course, this is not to say that good ideas did not emerge from this rather ambitious program. However, these good ideas, for want of a stabilized, hierarchically structured framework, will, from a standpoint of reaching modes of corrective action, more than likely have much the same effect as pouring a few gallons of water on a dry lake bed.

In an OMEGA matrix presentation, every effort would be made to offer a **hierarchically legitimate** structure for the above "spheres" and "issues". These would be indentured under not seven, but **only three** highest order "spheres"—the **Geosphere,** the **Biosphere** and the **Sociosphere**…More later.

OMEGA was derived on rhetorical, graphical and arithmetical grounds. It represents a hierarchically structuresd synthesis of all the controlling parts and issues of the global enterprise. As will be described in the next chapter, OMEGA is operationally expressed in three dimensions as a massive, fully integrated, entropy law-premised, Input-Process-Output-Feedback model. While also highly responsive to the need for complex mathematical modeling, OMEGA is designed so that valid operational decisions can be made with mathematics and micro-thinking in subordinate roles.

In the process of accommodating our world leaders in making truly sound decisions, we preclude premature involvement of the eventually higher mathematics necessary in complex problem resolution. Moreover, OMEGA could serve as a point of departure for persuading greater **operational relevancy** among the sciences steeped in higher mathematics—sciences, which history attests, often tend to proceed with great mathematical precision to answer the wrong question. Wrong because the question was not posed in the operational context of hierarchical order.

To belabor the point a bit further, let there always be right question asking in wholeness of context via the universal language of rhetoric, graphics and arithmetic…and not in the context of the mathematical maze which attends the electronic computerization aspects of a problem. Let it be underscored that **premature** electronic computerization and mathematical manipulation of inadequately formulated models has led many a decision maker astray.

Let us make full use of the capabilities of our growing information technology. But let us assure ourselves that the computer manipulation of any system or issue take place in response to a holistic model of the problem under study as lifted out of the wholeness of the OMEGA matrix. This will yield decision making information which is equally knowable to **all participants.**

Again, a truly highest order totality statement, in a pragmatic operational problem sense, is not mathematical. In any high order multi-variables problem, mathematics becomes a proper tool only after maturation of the comprehensive rhetorical-graphical-arithmetic statement. With now emerging twenty-first century problems of enormous scale, we must make every effort to assure that interdisciplinary communication has truly taken place at the nonmathematical levels of ethical overview, system validity and political sanction.

At this point in our discussion of the rationale for OMEGA, having premised certain of the basic conditions precedent to actual construction of the matrix, we will dwell, perhaps inordinately to some, on one of the most vital and least acknowledged aspects of the rationale for OMEGA. This is the precept of **analytical threeness** as it bears on stabilizing the hierarchical order of things. The precept of analytical threeness, as seen by this writer, since the technique is not yet subscribed to by academia, is a unique approach. It came about through an effort to offer maximum clarity in presenting large scale technological problems to nontechnical decision makers.

It is offered that those investigations involving heavy interdisciplinary collaboration on highest level analytical systems such as OMEGA should be hierarchically structured in no more than three parts, and in successive subordinate tiers of no more than three parts each. However, that level of OMEGA below which analytical threeness need not apply will be discussed during the actual construction of OMEGA in the next chapter.

Some analysts will tend to treat this vital precept too lightly at first, while others may well rejoice in a "blinding glimpse of the obvious". Again, bear in mind we are seeking clarity for the nonscientific high level decision maker...and in a way which also benefits the scientific analyst, whether or not he readily admits to it.

To begin with, we must recall to the conscious mind the significance of threeness in the structure of religion, science, government, symbology, communications, information processing, etc., etc. The threeness precept is particularly vital since the entire construction of the OMEGA matrix is premised on the concept of hierarchical order structured in a framework of analytical threeness. Once established, such an order of things may well prove virtually as unassailable as the multiplication table. Yet, how little use is made of this elegantly simple truth. A beautiful natural law of analysis as it applies to the whole of the global enterprise is available for the taking. But the utility of threeness is not widely evidenced in the conscious mind. There seems to persist an unspoken consensus to the effect

that in the hypercomplex twenty-first century, man cannot face the totality of his economic growth and ecosystem imbalance dilemma with simple structures of thought.

We must reject this festering consensus of thought. Analytical threeness, as the root totality of OMEGA, is a warranted assertion. Among
other things, threeness best facilitates the engendering of persuasive
ethical imagery on any single issue or integrated aggregate of issues. The
behavioral psychologists offer a simple basic supporting point for this.
They point out that in a wide array of considerations in a presentation,
people generally tend to remember the first item, the last item and the
item in the middle…threeness. A bit simplistic, yes. But it works.

Consider just a few of the more obvious threenesses and the great
offices and natural science platforms stemming therefrom. We find
threeness permeating our highest sense of the order of things, albeit
lodged, for the most part, in the subconscious. We find threeness in the
Holy Trinity. We find threeness in the Celestial Hierarchy. We find
threeness in the three prime elements of our immediate solar system—
sun, earth, moon. The three basic media of our immediate environment—land, water, atmosphere. The three basic branches of our
government—executive, legislative, judicial. The three phases indigenous to all information and computer modeling processes—input,
process, output. And a virtually endless further array of both symbolic
and highly utilitarian threenesses.

However, we shall not attempt to span the vast array of threenesses
found ad infinitum in both generic earth and the symbolisms of man.
One need only open the dictionary and scan the "tri" prefixes for
unlimited cues. In generic earth, the trimorphism of crystalography, the
trifolium of plant life and the trimolecularism of certain elements are
but a few. The symbolisms of man range from the triskeilion of early
Greece to a threeness somewhat in parallel with the structure of this
book—a trilogy of literature wherein three related plays each have their
own unity, but together form a larger work.

The whole point here is simply one of embracing a natural law of balance to be found in delineating the body of Principle discussed in Part I, the Purpose discussed in Part II and the Precept discussed in Part III in terms of a hierarchically ordered threeness of method—Principle, Purpose and Precept.

Our summary discussion of threeness would not be complete, of course, without noting the obvious—the presence of threeness in our every moment on this earth. Wherever that might be at any given time, it is continuously and unvaryingly executed in terms of three dimensionalness. So subtle in its implications, it seems simplistic at first to be reminded that our three dimensional personal bodies are located in three dimensions of space, the physical environments which serve our personal bodies are located in three dimensions of space occupying land, water and atmosphere, and, finally: Our three dimensional bodies and our three dimensional environments are harmonized with one another through the dynamics of operational process flowing through three levels of time cognizance—past, present and future.

Now, it is fully appreciated that this business of threeness may at this point sound all too obvious. To some it may appear that inordinate emphasis has been placed on analytical threeness as the approach to a basic method for dealing with oneness of humane purpose. But it is important to note that in today's analytical practices, there is a lack of clarity in communicating results...In presentations to groups of diverse disciplines, particularly political decision makers, the general tendency thus far has not been toward searching out simpler analogues of our massive problems. The result? All too often either there is no actionable resolution or a poor one.

Yet, threeness, and hierearchically subordinated multiples thereof, synthesized through the dynamics of operational process in flowing time, stands ready and waiting to yield a massive, yet elegantly simple analogue of the entire global enterprise. Moreover, it can do this in terms which are not only clearly expressed, but in terms which are clearly communicable across diverse professional disciplines, educational processes at all levels and the informed public...Indeed, in the precept of analytical threeness, working in concert with hierarchical order, there is found great beauty, elegant simplicity and intellectual serenity.

As we approach the conclusion of this chapter, we have discussed over the seven chapters thus far a wide array of issues and objectives in support of an ideal—humane purpose. But that eminently correct and lofty ideal has long suffered for want of pragmatic purchase on the hard and palpable realities pursuant to that ideal. When we address the pragmatics of overcoming inefficient operational systems in real world time through the OMEGA matrix, or a system akin to it, we have entered the path toward ecosystem balance. We can thus provide hard purchase on those lofty ideals which the professionals in the humanities have long urged but failed to implement.

Responding to the key issues outlined in the prologue, here are brief one line summaries of supporting objectives as discussed throughout the book:

To subordinate ethical indifference to ethical prerogatives.

To search out and respond to intuitional inferences.

To deny Techno-Anarchy as a self-directed economic catalyst.

To subordinate micro-specialism to macro-generalism.

To respond to the ecosystem countdown per Figure One.

To conduct growth analysis per the dynamics of Figure Two.

To structure an approach to differentiated economic growth.

To subordinate nanosecond computerism to natural time rhythms.

To advocate a formal entropic charter for interdisciplinarians.

To approach operational systems in time compression.

So, on to Chapter Eight and the construction of the OMEGA matrix in the context of a legitimate holistic structure for responding to the foregoing key issues and objectives.

*   *   *

CHAPTER EIGHT

Construction of The OMEGA Matrix

Actual construction of the matrix will take the form of a three-dimensional cube in order to obtain maximum real world visibility of the basic components of our analogue of the world structure. This means that analyses, conferences and dialogues on issues and solutions will always be conducted in a common language forum. That is, they will be supported by, but not dominated by, the language of Computopia. Also, while the format used for the matrix would be referred to by the technical analysts as a three-dimensional matrix calculus, we will not refer to it in that rather esoteric term.

Our reference to the matrix cube will be in the context of a practical, three-dimensional depiction of nearest neighbor interactions…So, let us begin. We are going to construct the OMEGA cube inward toward three major sections and twenty-seven OMEGA minors and their respective sub-minors. Any individual problem will always be seen in its relationship to the whole of things.

Hence, we approach the earth hierarchically, first viewing the inorganic **Geosphere** with its three prime elements—land, water and atmosphere. This forms our first slice through the whole cube…Then we add the second slice, the organic **Biosphere** with its three prime elements—people, animals and vegetation…Finally, we add the third slice through the cube, the **Sociosphere** with its three prime elements—urban systems, regional systems and world systems.

As previously noted, Geosphere-Biosphere-Sociosphere are those three hierarchically ranked components which form our ultrastructure axis of the whole of things. As will be shown as we describe the whole of the matrix, this is the output axis to be controlled by the entropy law.

Now, bear in mind that we are developing the structure of the OMEGA matrix in such a way that it will never change. However, the reader is urged, in the light of the mania for change we live under, not to jump to the wrong conclusion and object to the term "never". That is, numbers within OMEGA will come and go, and changes within the components of the matrix will go on ad infinitum. But the component definitions and their respective positions in the matrix will not change. The total OMEGA matrix is offered as a device for continuity of high level management purview of the world's responsiveness to the entropy law. The OMEGA format, conceivably, could prove virtually as unchanging and absolute as the multiplication table.

This is first seen in our description of the major divisions of the cube. There will always be a Geosphere, a Biosphere and a Sociosphere. By definition, they will always exist in the hierarchical order described. Likewise, the soon-to-be-described sub-cubes of the matrix will enjoy a comparable unchanging stability in their relationships to the whole of the matrix.

But before going further with the construction of the OMEGA cube proper, we need to view a two-dimensional reference index of complex first, second and third tier hierarchically ordered indentures. For clarity and easy return reference, we shall do this in the framework of the classical Input-Process-Output model, geared to the entropy law. This two-dimensional version is shown on Figure Three A.

Input is thought of as the Infrastructure, comprised of the three highest order elements of that axis—(1) People systems, (2) Commerce Systems and (3) Government Systems. People have always existed in response to some form of that highest order infrastructure. At the input level, all of the parts of the global enterprise are subordinate to one of the three elements of that axis.

Process is considered the energy conversion Suprastructure, comprised of the three highest order elements of that axis—(1) Energy Conversion Resource Usage, (2) Entropy Rate from Resource Usage, (3) Resulting Ecosystem Status. These three elements form the central core of OMEGA's response to the entropy law. Since energy conversion underlies **all** of the operational processes of mankind, we note that the use of natural resources in the energy conversion process produces an entropy rate. In turn, energy converted plus entropy rate is the root measure of our ecosystem status in terms of resources lost and attending pollution. Hence, all energy axis issues are subordinate to the analytical flow of those three elements.

Output, finally, is thought of as the Ultrastructure, comprised of the (1) Geosphere with its inorganic systems, (2) the Biosphere with its organic systems and (3) the Sociosphere with its urban systems, regional systems and world systems. As previously noted, those three elements of the matrix are the highest order outputs of the global enterprise and are the qualifiers of mankind's conduct of life in the context of the entropy law.

Parenthetically, there is obviously a feedback aspect to the above outlined classical analytical model framework. The feedback aspect is discussed extensively in the next and final chapter.

Now, the Input-Process-Output framework as illustrated is primarily a convenience from the standpoint of providing a relatively simple index of complex first, second and third tier hierarchically ordered indentures. However, in that two-dimensional depiction of the model in Figure Three A., interactions of one element with another are not readily seen nor manipulated. In addressing this consideration, then, we must perceive the operational earth sphere, with its randomly distributed elements, and note the need to grapple with the whole of things in a three-dimensional, hierarchically ordered frame of reference as earlier introduced on Figure 3.

To serve that end, we begin by assigning the previously described Input, Process and Output axes of the OMEGA matrix—Infrastructure, Suprastructure and Ultrastructure as indexed on the two-dimensional Figure Three A.—to a three-dimensional cube of those axes as shown on Figure Three B. In this three-dimensional juxtaposition of the parts of OMEGA, one may treat each axis in its own right, or at will examine any combination of elements of one axis with elements of another. That is, insofar as unique studies of one element or another are **always** addressed to the appropriate higher level in the total hierarchy. In this regard, it is underscored that the analyst will find himself unavoidably reckoning with the entropy law, functioning at the central core of the OMEGA matrix.

The structure of the whole cube is defined as OMEGA Major, comprised of twenty-seven level-two OMEGA Minors. In turn, each OMEGA Minor is comprised of twenty-seven third level OMEGA Sub-Minors. Each of the OMEGA Minors and Sub-Minors interplay with one another, depending on the depth of a particular analysis. Eventually, in the course of the analysis, they process upward through the hierarchy of OMEGA Major to find validation, rejection or corrective modification of the parameters of analysis.

In the course of studying Figures Three A. and B., one should note that for ease of cross reference, the indentured numbering system applies to both figures. Also, the reader is again encouraged to complete the whole text before suggesting the need for additional elements to the matrix.

To further illustrate the OMEGA construction process shown on Figure Three B., consider the hierarchical flow of interactive analyses as they relate to the infrastructure. For example, let us pull out OMEGA Minor 1.1x2.1x3.1—PeoplexEnergyxGeosphere.

Oversimply, people demand energy and dig into the Geosphere to get it—oil, coal, gas, whatever. But it is not enough to embrace a demand situation circumspectly. In any invasion of the inorganic Geosphere, it impacts in some way—in hierarchical order—on the organic Biosphere and, in turn, on the Sociosphere. Moreover, when people dig into the Geosphere to satisfy an energy demand, there will also be an impact on that particular OMEGA Minor of the suprastructure dealing with entropy and on the OMEGA Minor dealing with the ecosystem. Again, energy converted plus entropy rate equals ecosystem status.

Then, the interactive considerations triggered by the first OMEGA Minor, People, also impact on the OMEGA Minor addressed to Commerce and on the OMEGA Minor addressed to Government, thus comprising the interrogation of the whole of the Infrastructure axis at the Geosphere level. In turn, the Commerce and Government OMEGA Minors are also processed across the whole of the Suprastructure axis and, ultimately, across the whole of the Ultrastructure axis.

Now, to return briefly to the first OMEGA Minor, People. Consider the first of its Sub-Minors, Health , as illustrated. The interrogation procedure for this Sub-Minor as well as the other third level Sub-Minors under People is the same as described above.

Now, no one will argue that from a standpoint of the hierarchical order of things that People Systems is the appropriate first OMEGA Minor category of the Infrastructure axis. However, some will at first challenge the Sub-Minor delineations under People Systems—Health Systems, Habitat Systems and Education/Occupation Systems—as not sufficiently all embracing. Consider, then, this brief rationale.

Individually or collectively, if man has physical and mental health, a habitat to shelter himself and the education to occupy himself in the secular world with economic and spiritual equity, he then has all. Whether or not a man has health, habitat and education/occupation, any and all problems concerning people are found in those three Sub-Minor elements and **only** those three elements. This means, of course, that all other considerations below the third level are subsumed by the particular Sub-Minor being analyzed; with lower levels surviving hierarchical summation strictly in terms of validity and relevance.

Here again, it is a philosophy of control of carefully circumscribed analysis in a hierarchically ordered analytical threeness structure that we are subscribing to. Third level Sub-Minor elements not only retain that highly utilitarian symmetry of threeness requisite for mental efficiency and sound interdisciplinary activity, but enhance the precision of the management decision process in reckoning with interactive OMEGA components.

We can now note that the Sub-Minor level is the lowest level to which the precept of analytical threeness remains inviolate. Lower order indentures below the third level exist ad infinitum, of course. But, fortunately, the lower levels need not be held to three-dimensional delineations. That is, the lower levels will survive the rigors of OMEGA premised analysis only insofar as they can be legitimately processed and integrated through the upper threeness levels to which they are subordinate. In short, no lower order analysis need ever drift into that great void of fragmented nondirectional information garbage. The discipline of hierarchical relevance to be found in the OMEGA matrix can preclude that.

At this point, and for purposes of this book, we need not repeat the foregoing procedural approach to all twenty-seven OMEGA Minors and their respective Sub-Minors. The procedure is the same as that already described and applies to each of the components and elements of the matrix. However, any given area of particular interest may be analyzed at will as long as it passes the test of relevance and responsiveness to hierarchical order.

Having now described the construction of the matrix and its analytical flow process, a few words are in order as to the subsuming of certain indentures below the third level. Where the third tier hierarchical categories of OMEGA Minor number two under Commerce Systems are fairly obvious as to the threeness structure, consider OMEGA Minor number three under Government Systems as a less obvious example of analytical threeness in hierarchical order. This is mentioned because in lecture work the writer has invariably encountered debate on the concept of only three categories of government on the third level and further debate on the categories selected as insufficient delineation of Government Systems...

- **Ultrastructure Axis—Input**
  - Government Systems
    - Domestic Systems
    - Geopolitical Systems
    - Monetary Systems

Bearing in mind that we are reaching for a basic OMEGA structure which will not change down through the third level, consider the following rationale: It is speculated that while the indentures under Government Systems will always be quite large, Geopolitical Systems will be broken out as a separate third level OMEGA matrix entry reporting independently to the administration. It is further speculated that Geopolitical Systems , over the next half-century, will not only break away from the present Domestic Systems structure, but will grow vastly in size and importance as true globalization comes of age, notwithstanding foot dragging by predatory capitalism.

While this writer joins many other commentators in resistance to big government, it is offered that a special need exists in this case if a successful international consortium for ecosystem control is to be implemented. Geopolitical Systems, in addition to the Secretary of State, our ambassadorial functions and United Nations liaison, would also embrace within its structure a mirror image of certain Domestic System functions such as agriculture, ecology, energy and others. While a net increase in the size of government functions might take place, at the same time ineffectual international efforts toward ecosystem control, now buried in existing domestic systems, would quite likely flake off.

The government of every country responding to the OMEGA matrix—or to the inevitability of some kind of system closely akin to it—will have its own domestic systems, unique to its indigenous culture and the economic growth precedents it embraces. But in a climate of maturing globalization of economic growth and ecosystem control, they will be forced to also embrace powerful geopolitical structures corresponding to OMEGA or an OMEGA-like system.

The primary reason for the Infrastructure's third tier hierarchy as delineated—Domestic Systems, Geopolitical Systems, Monetary Systems—is to engender sharply focused effort on global economic growth and ecosystem control at the highest level. Indeed, so goes the ultimate global ecosystem, so goes the planet. Those three highest order categories under Government Systems, responding in concert to the OMEGA Matrix, invite the most balanced view of the global enterprise.

United States domestic systems breed a unique form of ecosystem imbalance with an attending unique management control situation. And every economic and ecosystem problem outside the contiguous United States is a unique geopolitical system problem of a different kind, more often than not requiring entirely different handling of a common problem—global ecosystem control. But in any case, the basic dimensions of OMEGA apply. Moreover, domestic systems and geopolitical systems problems invariably converge on monetary systems ground as the platform for reconciling whatever actions might be agreed upon…In short, the shifting sands of organizational change can come and go as long as they are subsumed under the **unchanging** basic structure of OMEGA.

Now, the above introductory discussion is not to suggest the impossible task of performing a major overhaul on the structure of the United States government or of private enterprise. Not that such is not needed in order to best serve the timeless beauty of the Constitution of the U.S.A. But that's another story. Rather, it is to underscore that administrative actions at the highest level first be operationally filtered through the hierarchically ordered, ecosystem-premised factors of OMEGA.

Since we seek essential structural containment of the virtually infinite variables which must interact to effect a sound economy and, in turn, a sound global ecosystem, we must acknowledge the practical constraints imposed on the OMEGA matrix concept. Half a century or more may well roll by before an essentially pure OMEGA becomes ingrained in the world's view of economic growth. However, it must be offered that highly useful approximate answers to the right questions at the right time will early on be found very much in evidence in the totality of the OMEGA way of thinking.

For purposes of this book, the whole structure as specified on Figure Three A. and Figure Three B. is offered as sufficiently self evident to further clarify the basic points of OMEGA construction...and to assure the idea of viewing the whole of things in terms of a full perspective of relevant interrelationships and processes.

That is just what OMEGA is: a relevance construction, always viewing a given situation as influencing or being influenced by those hierarchically structured neighboring factors impacting upon it. This means **graphical depiction** for decision-making clarity, and with necessary complex mathematical manipulation **rhetorically translated** to further augment the precision of the high level decision-making process...Clarity. Communicability. An action based, interdisciplinary language for reckoning with the imperatives of global ecosystem control.

*   *   *

CHAPTER NINE

Feedback From The OMEGA Matrix

Entering the final chapter of this book, lest the banner of idealism appear to wave too freely, it is acknowledged that early cycling of the OMEGA Input-Process-Output-Feedback model will be disappointing. As has been discussed throughout the book, there is much to do before acceptable OMEGA output and feedback data are realized. On the other hand, one of the prime objectives of OMEGA is to provide focus on the problem areas and provide guidance to paths of corrective action. Moreover, it is not a moment too soon to enter vigorously into implantation of the OMEGA way of thinking in the education system and in the professional community at large.

The geopolitical scene today is one that is heavily vested in futile efforts to bomb the world into submission. Highly publicized areas of conflict, together with many less publicized pockets of aggression, finds that reaching the world's greatest needs will come slowly. The opening of the prologue outlined the key issues involved. With the above in mind, this chapter will simply delineate the OMEGA model flow as a procedure, with emphasis on feedback.

Recall first, then, that OMEGA is a primal Input-Process Output-Feedback model wherein values associated with People Systems, Commerce Systems and Government Systems are seen **continuously** in their interactions with Energy Conversion, Entropy Rate and Ecosystem Status, thus value balancing, through feedback, the outputs of the Geosphere, the Biosphere, and the Sociosphere. See Figure Three C.,

The OMEGA approach to resolution of all of this would be formally implanted in the education system at large and taught in the context of time compression. As discussed, the concept of time compression addresses the ever increasing number of events to be contained within an inflexible period of time. It deals with the now limited units of time required to achieve harmonious dynamic integration of present as well as future operational systems with the grand total system comprising the global enterprise.

Actually, it will be in the process of seeking to contain time compression that mutual understanding will finally emerge between the environmentalists and the industrialists. Mutual recognition of the chaos caused by these two long opposed factions will come to pass and unmanaged chaos no longer allowed to persist.

The environmentalists will finally come to realize that explosive world wide economic growth is not something to be restrained, but rather a condition to be endorsed and lived with in harmony with the whole of things.

The industrialists, in turn, will also awaken to the reality that irresponsible economic growth would eventually terminate the flight of the planet earth through time and space. That is, insofar as the survival of the planet's ten-billion passengers is concerned.

Those two factions will eventually come together in a harmonious blend of ecosystem-sensitive operationalism, constitutional principle and ethics…and, in this context, will inescapably reckon with constitutional origins.

Surprisingly to the environmentalists, and particularly to the industrialists, it will become clearly evident that in the context of one environment, not only will humane purpose experience more focused response, but new dimensions of corporate profit will be unfolded.

Before reviewing the flow process of the OMEGA model, it can be noted, in a manner of speaking, that the present structure of economic growth management is a Rubic Cube wherein all the sub-cubes of the total cube are present but they are not juxtaposed or taught in their cor-rect relationship to one another. OMEGA optimizes the cube. So let us briefly review how the whole OMEGA system works from the top of the hierarchy to the ultimate output and feedback loop.

First, at the apex of the hierarchy, humane purpose: To establish health, habitat and education for the peoples of the world to a point of physical and spiritual equity for all…and from this…

Recognize the unprecedented explosive growth curves of the last half of the twentieth century, implying the never-to-end fierce growth of the world economy…while…

Responding to this growth in the context of time compression—the ever increasing number of events to be contained within a forever fixed number of units of time in real world time flow…Then…

Respond to the highest order problem confronting mankind in the con-text of the above, which is: The incipient implosion of the global ecosystem as triggered by a massive runaway entropy, world wide…Then…

Recognize the role of the OMEGA matrix, which offers a controlling system for managing runaway entropy, structuring a presiding para-digm for the entropy law…Thus…

Proceeding into OMEGA proper, the matrix receives **Input** demands across the entire Infrastructure of people systems, commerce systems and government systems. This input is then analyzed across…

The Suprastructure, the **Process** axis of OMEGA wherein a given unit of energy is not responded to in isolation, but rather is challenged in terms of its entropy rate and its resulting ecosystem effect in the con-text of the total breadth of the world's energy conversion demands.

Parenthetically, the entropy law is the operational nexus of the OMEGA matrix hierarchical structure. The ever increasing intensity of the energy conversion processes going on in the world today, when addressed in any kind of summary, are done so in an alarmingly fragmented, uncoordi-nated fashion…The result? The previously noted incipient implosion of the global ecosystem…So, with this in mind, we can interrogate…

The Ultrastructure, the **Output** axis of OMEGA. In the context of nearest neighbor interactions , as the **Input** demands of the Infrastructure undergo **Process** through the energy axis Suprastructure to the **Output** of the Ultrastructure's Geosphere, Biosphere and Sociosphere, every energy conversion process is interrogated in terms of its entropy rate and its collective impact on the global ecosystem…Consider an oversimplified illustration.

Man responds to an energy conversion demand and digs a hole in the Geosphere—inorganic earth. This act, in turn, impacts in some way on the Biosphere—organic earth. In turn, this energy conversion act impacts on the Sociosphere—the whole of world society. This produces a net effect on the ecosystem which either is or is not within acceptable limits…acceptable, that is, as it bears on establishing and maintaining the quality of the global ecosystem…

Finally, with meaningful **Feedback** data from the sociosphere to present to a "World Council on Ecosystem Control", findings are processed through constitutional principle and ethical parameters. These data are then analyzed in terms of response to the goal of humane purpose at the top of the hierarchy…Thus…

The cycle is complete, but continuing, as this now analyzed Feedback is reentered into the demand Input of the OMEGA Infrastructure and recycled periodically.

In summary, and this bears repeating, the OMEGA matrix can offer an orderly hierarchical progression from Infrastructure to Suprastructure to Ultrastructure. That is, from the **Input** of People, Commerce and Government to the **Process** of Energy, Entropy and Ecosystem to the collective **Output** of the Geosphere, Biosphere and Sociosphere. All that while always bearing in mind the entropy law and the relentless time compression thrust upon us by the never-to-end explosive growth of the global enterprise.

The outcome of all this?…An eventual stable global ecosystem dedicated to humane purpose. Thus the planet's ten-billion passengers and their successors will be found in safe passage.

In a few closing remarks of this last chapter preceding the epilogue, the writer acknowledges that this book could easily have run to a one-thousand page text rather than to its relatively concise one-hundred-fifty pages. But in presuming to address so wide a spectrum of subjects and key issues, with emphasis on the imperatives of hierarchical order, it was decided not to develop a large, inescapably ponderous text. It was felt that in doing so, there would be a risk of losing the feel of the total OMEGA structure and the key issues it highlights.

Therefore, the informed reader is offered a loom upon which to weave the whole cloth. Of course, many of the myriad elements comprising OMEGA and the key issues have been addressed by other writers. Many of these excellent writings amplify the elements and issues presented herein.

In closing, here again is the nexus of OMEGA. It is, concomitantly, a device for orderly qualitative and quantitative measure and macro-management of the global whole of things as well as a platform for responding to: "One Nation, under God, indivisible, with liberty and justice-for all"…Among many provisions related to this, the founders of the constitution put forth the foundation of "E Pluribus Unum". They stated:

"We the people of the United States, in order to form a more perfect union, establish justice, insure domestic tranquility, provide for the common defense, promote the general welfare, and secure the blessings of liberty to ourselves and our posterity, do ordain and establish this Constitution for the United States of America."

However, in the course of-full endorsement of the founders' timeless statement, "We the People" must become fully aware of a caveat which could not have been foreseen by the founders of the constitution. In short, an antithetical oligarchy has been emerging through Techno-Anarchy, the many dimensions of which we have discussed in the text.

The technologists now press the politicians into info-games, often involving questionable computer manipulation of data, which leaves the decision-making process a house of mirrors. The distortions imposed on the decision-making process can only worsen as these mirrors reflect still additional mirrors arising out of an unchecked, rampant growth of information—information which has not been subjected to a barometer of relevance to "We the People".

However, if we reawaken to constitutional principle and timely ethical parameters, and then reinforce never-to-end economic growth with the concepts offered via the OMEGA matrix, and, in turn...

If we bring **the education community as a whole** to a capacity for wide dissemination of the OMEGA approach in the context of the constitution...

Then "We the People" will be able to walk a new and lasting bridge between the pillar of religion and the pillar of science in the context of world loyalty. From that vantage point, with idealism merged with true realism, we will be enabled to approach with enhanced authority the enormous task of putting ecological rudder to global enterprise.

        *   *   *

EPILOGUE

Parthenon West—The Year 2050—An Allegory

As he drove up the California coast toward Parthenon West in this year 2050, the Director of The OMEGA Institute contemplated the difficult but progressive decades he and his world-wide associates had endured. Their quest for domestic and global response to the actions needed for a balanced ecosystem dedicated to humane economic growth was now bearing substantial fruit.

But it was not always so.

Before the advent of Parthenon West, notwithstanding the much appreciated support of empathetic elements of the media, the OMEGA Institute found early on in the twenty-first century that some dramatic new approach was needed. Some new concept was necessary in order to generate increased general awareness of the need to greatly accelerate progress toward reckoning with the moral and ethical aspects of a global population rapidly growing to ten-billion people.

This led to the founding of Parthenon West as the economic core of the great new city which now surrounds it. The OMEGA Institute's principles and objectives were now dramatically couched for eliciting timely response and action.

What, then, was so different?…Actually, before Parthenon West came into being, an approach to the proper handling of a somewhat elusive macro-communications problem came to the director during a casual vacation trip to Greece and the Parthenon of the Acropolis. To the receptive heart, there was no mistaking the lingering spiritual atmosphere and the great intrinsic beauty of the massively scaled structure. Though the Parthenon housed only the Godess Athena—"Wise in the arts of war and the industries of peace"—the symbol was world renowned.

Since his goal was to discover a medium for achieving vastly intensified awareness of the global problem, the director asked himself if an architectural symbol of great beauty and import could be designed which would lift vitally important holistic thought out of the doldrums of present communication media. The idea of an institute structure on a grand scale could certainly address one aspect of the problem. But reaching the high level decision makers of government and industry on both a physical and spiritual basis, lifting them above a plague of fragmented thinking, away from political game playing and into a high level of moral and ethical response, would require much more than architectural beauty and grand scale per se.

As soon as he returned to his OMEGA Institute headquarters, the director assembled a study group comprised of top architects, theatrical producers, behavioral scientists, design artists and design engineers. His intuition was soon to be confirmed to the effect that an action-inspiring portrayal of the world condition would have to be dramatically cast in an ambience of grand theatre and great intrinsic beauty.

Beginning the group study with the question as to the portrayal of the world situation in the most dramatic and palpable manner, it was quickly concluded that ponderous written reports, two-dimensional maps, graphic displays found too sterile in a sea of static displays, and even well presented animated films of various ecosystem problems, were not getting the job done. These media were not generating sufficiently meaningful response from the power centers of decision making.

Responding, then, to cues from the behavioral scientists on their views of two-dimensional stimuli versus three-dimensional stimuli of various scales, it soon became clear that a massively scaled globe of the earth was the first order of study. The proper scale for the globe would stimulate the most hard nosed observer with the experience of an enlightening three-dimensional presence of the earth in response-inspiring awesome proportions.

The next step, of course, was to determine the optimum size of the globe. Guided by the spatial geometric thinking of the architects concerning viewing angles, viewing balcony locations and related architectural design considerations, they suggested, as a maximum practical design and construction size, a diameter of one-hundred feet.

The theatre dramatists, the design people and the OMEGA Institute analysts found this recommended size ideal. This would provide for both appropriately awesome drama and an adequate global surface space for computerized displays of all problem areas either simultaneously or individually as called for by the presentation moderator.

The next step was that of outlining tentative specifications of computerized earth surface displays. The interior of the globe would prove to be an unseen but very busy place with permanent access scaffoldings at various latitudes, a movable boom for access to northernmost latitudes and, central to it all, the surface display computer systems and operator control consoles. The central control console operator, upon call from the presentation moderator on the outside, would also operate earth surface display projectors of certain problem areas from projectors mounted in the facias of the viewing balconies. This dual capability of inside and outside surface display was to more readily accomodate the shifting tides of change. Certain displays would be lighted semi-permanent displays projected from the interior, others would not.

For example, the world's current areas of armed conflict, to the degree that they impact negatively on ecosystem considerations—and they all do—would be simultaneously displayed in red lights from the interior as, sadly, a viewing baseline for many ecosystem considerations. At the opening of the twenty-first century, there were over forty hot spots of armed conflict around the world and, according to the geopolitical experts, this number was expected to increase. At least this would be the case until such time as sufficient awareness of the breadth of the ecosystem and hunger war took place. Such awareness, of course, would eventually unite all ethnic cultures in an inescapable common cause.

Returning to the completion of the study group's preliminary planning on earth surface displays, they listed myriad other global surface displays ranging from agriculture, water and energy status to the demographics of population growth, migration and world hunger patterns…together with ecosystem problem areas and so forth. All in response to OMEGA matrix output data.

But there was a still further basic need. To somehow couple the awesome physical reality of massive scale with the spiritual insights of the founding fathers of America. The "In God We Trust" guiding rule stamped on our coinage, together with "E Pluribus Unum", should be freshly endorsed with a palpable quality of overwhelming beauty. It was decided that great architectural beauty and grace, coupled with symbolic sculptural entablature and commemorative bronze plaques would inspire further spiritual insight in reckoning with the gargantuan tasks facing us now and in the decades to come. Indeed, **the Bergsonian philosophy of the blending of artistic directness with science** would at last be realized.

At this point, with the carefully outlined preliminary design in hand, the project was a very real possibility if momentous financial obstacles could be overcome. First off, there could not be government financing of any kind. The concept could not be fettered by the pressures of of the bureaucratic quagmire, though it was intended that the government become very much involved in the resolution of issues after the project came into being.

The project would have to be privately subscribed. Following a year of further study and analysis, the OMEGA Institute had developed basic plans and taken an option on a one-thousand acre site. After putting down their own qualms over the unexpected multi-billion dollar cost of doing it right, the institute's officers spent another year with gratifying success. Not without intense effort, the OMEGA Institute developed a small group of very wealthy people who came to understand the significance of so vital a project and who, in turn, committed to the financing. As a practical incentive, since the project was to be a nonprofit foundation, these financial people were made keenly aware of the business potential of the great new city that was to surround the project.

"Alright", they said, "you now have your financing and your Board of Directors. But what do we name so important a project?"

With hardly a pause for breath, the director smiled. "We call the project Parthenon West! What else!"…

*   *   *

Still contemplating the evolution of Parthenon West, the director paused at its entrance pavilion. It was just after dawn on a clear new day, long before the staff and various conference groups would arrive. After chatting briefly with the security people, he decided to leave his car in the underground garage of the entrance pavilion rather than to take the service tunnel to the parking area underneath his office. He would walk the half-mile on the landscaped drive to the main entrance of Parthenon West.

With the beautiful grounds to himself on this tenth anniversary of the opening dedication of Parthenon West, he yielded his reflective mood to a quiet encompassing of the whole of that which many years of tremendous effort and perseverance had wrought. On this special anniversary, he would be sharing his thoughts with several hundred high ranking executives of government, industry, education and key research centers from all over the world at ceremonies to be held later this day.

Parthenon West embraces a one-thousand acre site on the California ocean front. Between the main highway entrance pavilion and the surf impacting on the rocky shoreline, the whole complex is contained by a peripheral drive one nautical mile in diameter. The drive encircles a large gentle knoll. At the center point of the circle stands the magnificent main structure of Parthenon West, which also houses the OMEGA Institute.

The main structure, also a perfect circle, is, with its sculptural entablature and symbolic detail delineation, a symphony of architectural beauty and grace. The one-thousand foot diameter main building is surrounded by four small conference retreat pavilions and the chapel, located one-quarter mile from the center on compass radials.

The purposeful symmetry of the architecture is given graceful asymmetry through its concentric landscape delineations. Walking paths around expansive lawns, gardens, trees, lakes, fountains and rock formations are further accented by heroic scaled field sculpture.

As he reached the main entrance, he climbed the seven steps to the exterior promenade and walked between the seven-story-high,tri-fluted, tapered columns spaced every ten degrees on the compass rose. Walking across the wide promenade which encircles the building, he entered the dramatic main foyer with its soaring ceiling, symbolic sculpture and fine paintings themed to the Parthenon West mission.

He walked straight through the foyer to the center of the amphitheater and viewed the massive one-hundred-foot diameter model of the earth floating on its base pedestal in a large two-hundred foot diameter hemispherical bowl with the equator at floor level. Computer operated dynamic earth surface displays, with their operator control consoles housed in the center of the globe, would come alive later. The seven story amphitheater was covered by a glass dome. On the full circumference of the dome, the flags of all nations were flown.

On the first two floors, the space between the circular viewing promenade around the globe and the outer wall of the building provides for the auditorium, lecture halls, conference rooms, communications center, banquet room, staff restaurant and visitor administration center.

The third, fourth and fifth floors housed the OMEGA Infrastructure, Suprastructure and Ultrastructure laboratories respectively.

The sixth and seventh floors housed the Parthenon West executive offices, staff and administrative functions.

The seventh floor also provides offices and services for longer term professionals-in-residence from outside agencies.

The director would share further details of the whole complex during the anniversary ceremonies later in the day. Though many of those who would be in attendance were already participants in Parthenon West programs, the director fully intended no let up in the further persuasion of advocates—those who would carry the message of success to an ever widening global audience. He would also extend further welcome to those companion institutes who needed to share Parthenon West's ability to more dramatically present their own findings.

Continuing his reflective stroll, the Director of the OMEGA Institute, who is also chief executive officer of Parthenon West, walked around the earth model's equator to the elevator and took himself to the sixth floor and his office suite. His office is centered on the west cardinal point of the compass, looking across the inspired landscape to the sea where the early veils of mist were already dispersing in the morning sunlight. Further contemplating the mind expanding beauty of all that lay before him, the director thought back to the stimulating and rewarding experience of working with the architects, artists and design engineers over a long period of time.

Since the key architects, artists and design engineers were to be present at this tenth anniversary celebration, he would honor them by sharing with the whole audience the circumstances which led to the magnificence of the Parthenon West complex.

　　　　　*　　　　　　*　　　　　　*

Many years ago, the Board of Directors agreed that Parthenon West was a project which would spare no legitimate expense in achieving the most dramatic possible blend of beauty, form and function. A top architectural firm was selected which evidenced a sound enough cerebral understanding of Parthenon West objectives, but the director was not satisfied. He wanted a piece of their soul.

The architectural firm groused a bit when the director insisted that the final negotiation of contract details take place in Greece. Why?...They had asked. They had his message. They were enthused over the project, too. Hadn't they been picked because they knew there way around in big project execution? Look at their record. Moreover, no first rate architectural firm in their business was without its architects who had the poetic feel of ancient Greece. All right, they agreed among themselves, let's humor him. He represents our largest and most exiting new project and we anticipate some pretty solid corporate promotional value upon completion of the project. So, if he wants to wrap up the deal in the so-called magic aura of the Parthenon of the Acropolis, let's go.

The director had no desire to appear mysterious in his request. He had been quite circumspect in confining his reason to wanting on-site discussion of some of the physical details of the structures of the Acropolis...But he had another purpose in mind. While confident that the firm selected was the right one, the philosophical thoughts he would share with them after a few days, when the mood was right, could well make the difference between a flamboyant execution of Parthenon West in a Las Vegas mode, and the rendering of Parthenon West as an enduring statement of the best of western culture of the twenty-first century.

The rendering of Parthenon West would represent the western world's first attempt to restate, in contempoary symbolism, the heritage of democratic government which in so many ways reflects the philosophies of ancient Greece.

Indeed, the rendering of Parthenon West was to reach a great deal further, combining symbolic statement with an internal fibre of purpose. Substantive work on the great problems of our time was to take place in the very shadow of the structural symbolism to which Parthenon West would be delineated.

Every year, established architects, attorneys and people of many professions, creeds and languages unabashedly made pilgramages to Greece, seeking in the ruins of the Parthenon a sense of heritage which had either slipped from their grasp or suffering for want of nurture. While the Acropolis was overrun during the summer season with young student aspirants to the professions, avidly collecting superficial experiences, there were others too. For every starry-eyed youngster, there was a battle scarred adult who had had it up to here in his daily contest with the professional rat race and was seeking a spiritual uplift.

Most interesting that this one spot in the world, and a collapsed culture at that, should have such drawing power. Professionals from all over the world simply hoped to find new inspiration in retracing the very ground walked on by Socrates…Phidias…Pericles…and other ancient advocates of democratic government in a free republic.

The OMEGA Institute group felt it imperative to induce the architectural people who would work with them in the execution of Parthenon West to feel the spirit that still lingered among the classic ruins of the Acropolis…What motivated Phidias, Pericles, Ictinus and Calicrates in their great collaboration on the Parthenon following the Persian wars? What motivated these collaborators to work so feverishly, yet making no compromise with precision in the execution and sculptural delineation of the Parthenon? What motivated Pericles to fight every step of the way over administrative obstacles posed by his political opponents? Considering the remarkable speed with which this great task was executed, they must have somehow sensed that the Athenian culture had attained its peak of maturity, spiritual depth and artistic sensibility…and that since the Athenians were already growing complacent, their society would one day collapse, leaving no significant statement of a once great culture.

In sharing this same sense of urgency more than two millennia later, the OMEGA director hoped to put down a spectre of collapse not unlike that sensed by Phidias and Pericles. Yet, should our own cultural heritage one day decay through complacency and moral malnutrition, the remains of Parthenon West would at least furnish future generations a platform for making new constructs...But he hoped for far more out of Parthenon West than the neo ruins of the twenty-first century.

It was during the excursions to the Acropolis each afternoon follow- ing the morning contract meetings that the lingering spirit of this city near the Aegean Sea began to have its impact on the architects who would render Parthenon West. Unlike their friendly indifference at the outset of the trip to Greece, they were now caught up in the spirit of the venture with no further prompting from the OMEGA group. Here was immediacy...The architects allowed themselves to become short of breath in their afternoon climbs in the cool crisp air of late autumn high above the Aegean Sea. They touched the weathered marble of the Parthenon, climbed the high steps of the crepidoma, examined the sculptural delineation, mentally scaled the Doric columns...Each morning they spent with the OMEGA group over the details of Parthenon West, each afternoon living with the spirit of the Parthenon of the Acropolis.

Concrescence was at hand. Two architectural modes—the Parthenon of the Acropolis and Parthenon West—were being seen as one timeless spirit. Separated by over twenty-four centuries insofar as architectural technique was concerned, but separated not at all in the quality of spirit that motivated the two works. Reflection of this adventure of the human spirit was felt in the morning contractual meetings as the top architects of the firm brought their latent selves to the surface. The quality of inspired design thinking that the director had hoped for was now most acceptably in evidence...He now had a piece of their soul.

The OMEGA director felt that the architects were now prepared in head and heart to listen to his yet unspoken true reason for bringing them to Greece. In the afternoon of the tenth day in Greece, he made a presentation to them; summing up philosophically and pragmatically the whole of what Parthenon West was to be all about...

*　　　　　*　　　　　*

"While the Parthenon of the Acropolis and Parthenon West are of one spirit", he said as he opened his presentation, "the gulf in the degrees of knowledge obtaining between the period of Phidias and the period of those of us who will execute Parthenon West imposes extra-contractual, philosophical considerations. Many of these considerations are not of a specification nature, but should be known and heartfelt by those who will direct major phases of the work. This presentation will highlight those considerations and, hopefully, relieve any tendency toward sterility or inconsistency in the delineation of the myriad details comprising the whole.

"Remember, too, that the Parthenon of the Acropolis contained only the Godess Athena, whereas Parthenon West will encompass a depiction of the entire global enterprise.

"The Parthenon we've been visiting for the past ten days was executed during a period where one man of Phidias' calibre could contain virtually the whole of knowledge at that time. No single department of thought was strange to Phidias. In addition to his depth in the philosophies, he was painter, sculptor and architect. Phidias dominated the arts and the humanities of the fifth century B.C. in a culmination of tenured Greek culture.

"Unlike the period of Phidias, where one man could contain essentially the whole of knowledge, or at least stand back to back with his contemporaries in just one circle of knowledge, Parthenon West will have to reckon with two basic circles of knowledge. The first of these resembles the Phidias-like circle of the arts and humanities—the non-physical sciences. The second circle of knowledge is the physical sciences and technology.

"As covered with you, the architects, in our earlier contractual discussions, we shall, through the architectural impetus of dramatic scale and beauty, coupled with heretofore unapplied interdisciplinary principles, engage the two circles of knowledge in a most unique and unprecedented approach to the great issues of our time. This sharing of thoughts with you today, while obviously a summary of much of that which is known to you, is, in its most important aspect, an offering which synthesizes ideas and values in a way that will prove new to you.

"Consider first the oldest of these two circles of knowledge—the community of the arts and humanities, and the nonphysical sciences. As you know, this classical community has grown through time to acquire highly authoritative perspectives on the approach to global issues and humane purpose. Yet, over recent decades, it has gradually yielded itself to offering little more than social comment. Under the manipulative pressures of Techno-Anarchy, it cannot be construed as a community given to innovating and executing new plans for social betterment. With rare exception, the nonphysical sciences community now settles for being a part of effect, not cause. While its wide range of meaningful comment often leads to new thinking, that community is very weak in implementation posture when compared with the second and politically dominant circle of knowledge.

"The physical sciences and technology represent that community whose systems are preponderantly subject to observation through the five physical senses. Permeating every aspect of our lives as it does, we have come to accept many modes of conduct by that community, more often than not failing to put forth warranted moral and ethical challenge of their actions. In their now significant tenure, the physical sciences and technology have aquired enormous political clout, suppressing the more warranted agendas of the nonphysical sciences.

"The main point we will challenge via Parthenon West is the thin response by the physical sciences and technology to moral and ethical issues bearing on humane purpose. In turn, we shall attempt to assuage its often devastatingly powerful unilateral role in cause and effect. Concomitantly, we shall seek to establish appropriate prerogatives for the nonphysical sciences.

"The nonphysical sciences, of course, include theology, philosophy, law, economics, politics, sociology, demographics, noetics and a host of other eddy currents of knowledge. Now, as should at this point be quite obvious, this circle of knowledge to a large degree suffers separation from the other. It has no organized interdisciplinary platform for exchange with the physical sciences. Yet, the nonphysical sciences represent the very backbone of the pursuit of moral and ethical issues.

"Bringing these two divergent communities together on ethical bases and reaching the minds of those who must carry the greatest macro-management burden mankind has yet known, will be one of the enduring tasks joined by Parthenon West. The point to be underscored concerning the interrelationship of the two circles of knowledge is that, yes, they do work together at times, but that work, with rare exception, is seriously fragmented, usually without adequate prioritization or response to the hierarchical order of things. It is in the interdisciplinary melding of these two circles that the great issues of our time will find emergence in their most purposeful and manageable form.

"Here with us today we have the directors of each of the two circles of knowledge. The OMEGA Institute has a staff comprised of representatives of all professions. They are managed in an interdisciplinary way to respond to our director of operations, also with us today. The operations director has the responsibility of assuring equitable response to the disciplines of both circles of knowledge in presenting the dramatic physical depictions of the many real world issues addressed by Parthenon West.

"It is in the dramatic presentation of a galaxy of global issues in clearly understandable, aesthetically inspired physical terms that we hope to bring about a convergence of the two circles of knowledge to a unified value system. While motivated by the spiritual aspect, the issues are expressed in terms of the physical aspects of human dignity in the belief that by so expressing them, understanding and response by the decision makers will be most rapidly attained. In turn, the two communities of knowledge will be more quickly persuaded to unite their efforts along lines oriented to physical and spiritual equity for all people.

"An OMEGA Institute associate put it most succinctly when he suggested that what we are really after is a 'Unified Field Theory' for the conduct of the global enterprise.

"Finally, gentlemen, on one hand all of this may sound like an ambitious and difficult scheme—which, actually, it is. But, on the other hand, it will at the very least focus the minds of men and women on the highest understandable system of **operational values** bearing on our globally oriented humane purpose.

"We conclude this brief presentation here in Athens on the note that the heart and precision with which this project is executed will determine to a very great extent the degree to which exploitation of Parthenon West is successful. Precision in this case is two-fold: One, design and craft precision, for which you are already highly regarded; and, two, philisophical precision of a unique character in the interpretation of major finishing details yet to be worked out. It is this second aspect of precision which caused us to invite you to Greece and to share with you that aspect of this task which does not lend itself to contractual language. It is our hope that both the practical and philosophical considerations offered in this presentation are carried to a maximum depth of heartfelt understanding among the hundreds of people working under our joint direction for the next several years...Thank you for your fine response."

*   *   *

On this early morning of the tenth anniversary of Parthenon West, the director reflected on his presentation to the architects in Athens many years ago; and on the glorious outcome of the great collaborative effort that brought the OMEGA Institute to this day. He would draw on the essence of that presentation in his talk during the anniversary ceremonies...With deep gratitude to God for His unerring guidance, he once again surveyed the sweep of Parthenon West's now timeless domain.

The director mused happily over the comment of one critic in particular. Upon its completion, a respected critic ascribed to Parthenon
West the same words Plutarch had offered in commenting on the completion of the Parthenon of the Acropolis: "Through its great beauty, it
became at once antique!"

*   *   *

The global' enterprise now has a rudder. World leaders have manned
the helm and, with humane purpose, are steering diligently toward containment of the earth's ten-billion passengers.

The bridge between the pillar of the spirit of religion and the pillar of
the spirit of science in the context of world loyalty has been con-
structed. That bridge is now being walked in both directions.

*   *   *

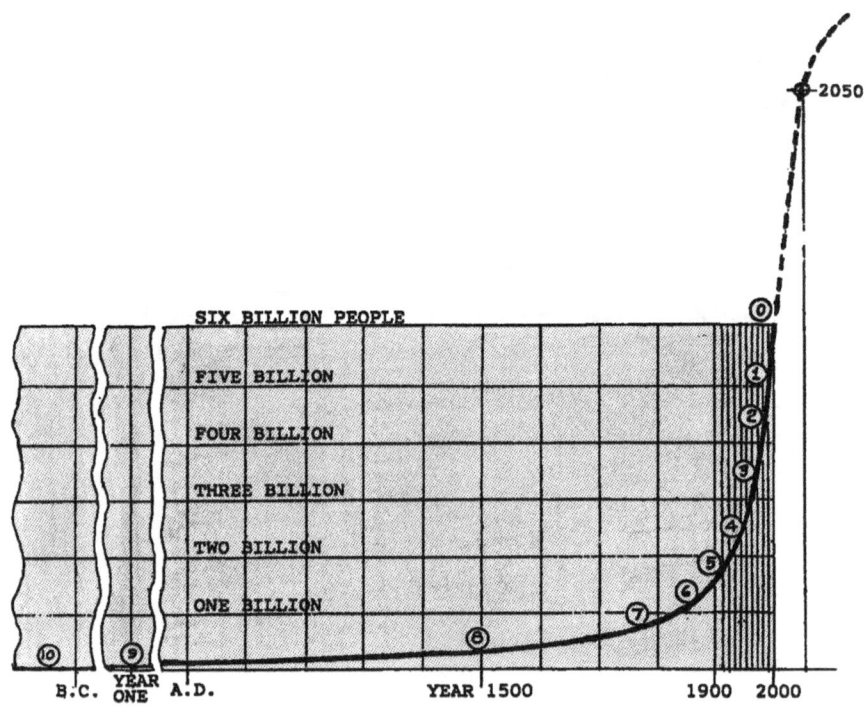

Figure 1.

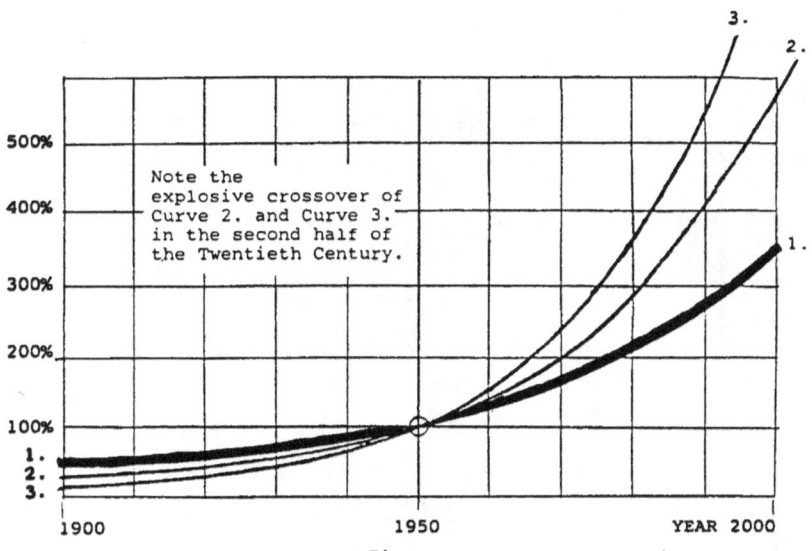

Figure 2.

Omni Matrix, Economic Growth Analysis
(In Clsssical INPUT-PROCESS-OUTPUT Model Framework)

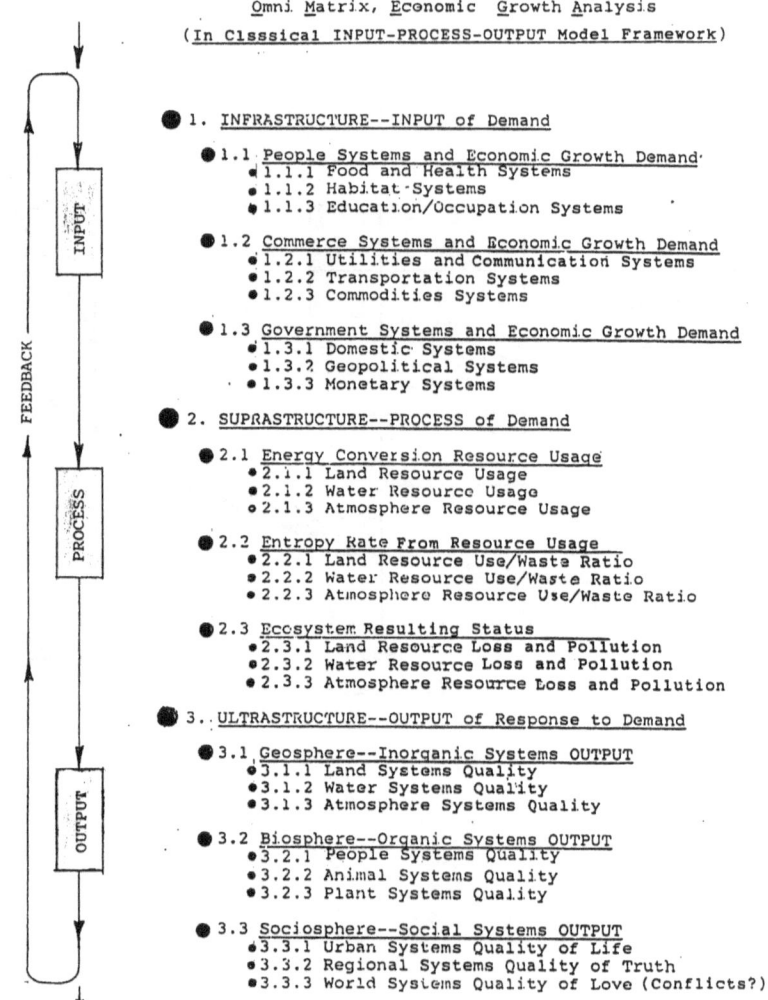

1. INFRASTRUCTURE--INPUT of Demand

    1.1 People Systems and Economic Growth Demand
        1.1.1 Food and Health Systems
        1.1.2 Habitat Systems
        1.1.3 Education/Occupation Systems

    1.2 Commerce Systems and Economic Growth Demand
        1.2.1 Utilities and Communication Systems
        1.2.2 Transportation Systems
        1.2.3 Commodities Systems

    1.3 Government Systems and Economic Growth Demand
        1.3.1 Domestic Systems
        1.3.2 Geopolitical Systems
        1.3.3 Monetary Systems

2. SUPRASTRUCTURE--PROCESS of Demand

    2.1 Energy Conversion Resource Usage
        2.1.1 Land Resource Usage
        2.1.2 Water Resource Usage
        2.1.3 Atmosphere Resource Usage

    2.2 Entropy Rate From Resource Usage
        2.2.1 Land Resource Use/Waste Ratio
        2.2.2 Water Resource Use/Waste Ratio
        2.2.3 Atmosphere Resource Use/Waste Ratio

    2.3 Ecosystem Resulting Status
        2.3.1 Land Resource Loss and Pollution
        2.3.2 Water Resource Loss and Pollution
        2.3.3 Atmosphere Resource Loss and Pollution

3. ULTRASTRUCTURE--OUTPUT of Response to Demand

    3.1 Geosphere--Inorganic Systems OUTPUT
        3.1.1 Land Systems Quality
        3.1.2 Water Systems Quality
        3.1.3 Atmosphere Systems Quality

    3.2 Biosphere--Organic Systems OUTPUT
        3.2.1 People Systems Quality
        3.2.2 Animal Systems Quality
        3.2.3 Plant Systems Quality

    3.3 Sociosphere--Social Systems OUTPUT
        3.3.1 Urban Systems Quality of Life
        3.3.2 Regional Systems Quality of Truth
        3.3.3 World Systems Quality of Love (Conflicts?)

Figure 3.A.

Figure 3.B.

THE OMEGA MATRIX

1. INFRASTRUCTURE Axis—INPUT

2. SUPRASTRUCTURE Axis—PROCESS

3. ULTRASTRUCTURE Axis—OUTPUT

   See Figure 3.A.
          for
      Correlation

   See Figure 3.C.
          for
      Matrix Flow

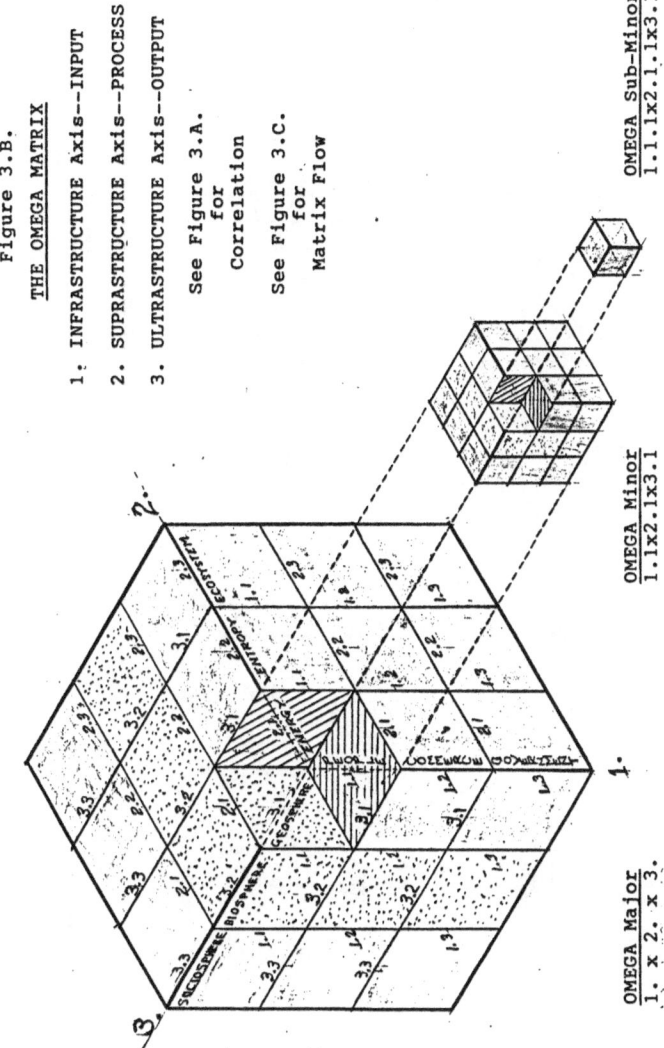

OMEGA Major
1. x 2. x 3.

OMEGA Minor
1.1x2.1x3.1

OMEGA Sub-Minor
1.1.1x2.1.1x3.1.

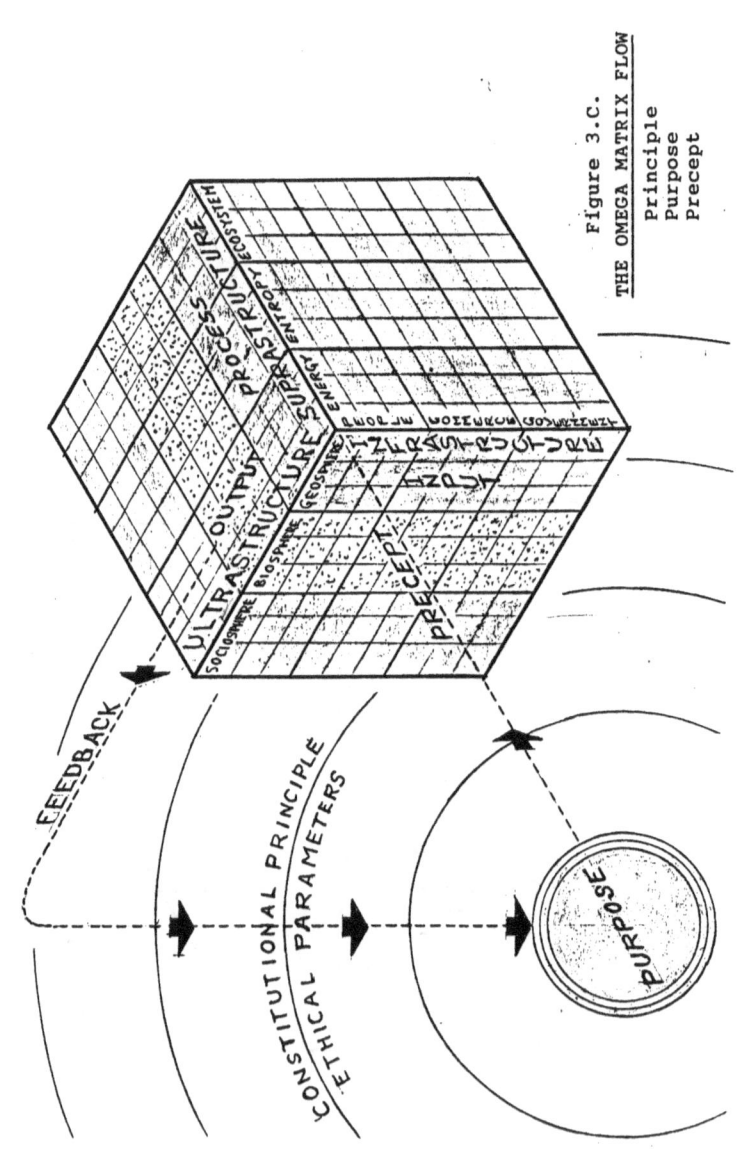

Figure 3.C.
THE OMEGA MATRIX FLOW

Principle
Purpose
Precept

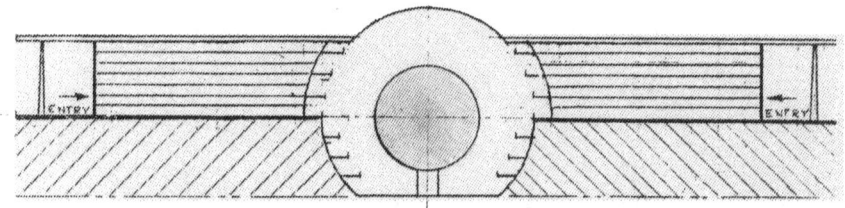

Figure 4.

Schematic Cross Section of Parthenon West

RALPH H. MINER BACKGROUND SUMMARY AND PROFESSIONAL ORIENTATION

Upon graduation in aeronautical engineering and commercial pilot training, I first accepted a position as a Junior Design Engineer with a major aerospace corporation. From earliest years, my professional inclination was to study and react to a wide gulf in modus operandi between the scientists and engineers who designed things and the operational people who must put these things to use in some purpose oriented system. Over the years, joining others who shared my views, my professional life has been aimed toward the development of design techniques and management systems for improving operational system efficiency and interdisciplinary integration.

My orientation led me early on in World War II to resign my position as an aeronautical engineer on the **design** side of the fence and to accept a direct commission as a Captain in the U.S. Air Force on the **operational** side of the fence. Assignments were as an international pilot-in-command in all types of fighter, bomber and transport aircraft ranging all over the world from the arctic circle of Greenland and Iceland to the equatorial regions of Africa and South America. My final war time assignment was as Chief Pilot and Director of Training of an Air Force base in Northern India, flying four-engine transport airplanes over the treacherous Himalayan mountains of Tibet to China. Many thoughts on new approaches to operational system design began to germinate during this period, not the least of which was an embryonic system view of the ecology and sociology of the many countries visited.

Returning to civilian life after the war, a number of executive positions were held in addition to my work as a Design Engineer and Operations

Research Analyst...Was General Manager of the Commercial Aircraft Marketing and Product Development Division of North American Aviation...Was Civilian Technical Consultant to the Deputy Inspector General Of the Air Force for Aircraft Industry Technical Inspection and Flight Safety...Was Division Director of five-hundred scientists and engineers for the Lockheed Missiles and Space Corporation on advanced state-of-the-art programs...Was Research and Development Consultant to the President of Lockheed Missiles and Space Co. This last assignment led to applying new analytical techniques to my direction of a ten-man interdisciplinary team in searching out ecologically premised new business opportunities which might exist on the then-distant horizon. However, after spending a very large amount of corporate research budget on this uniquely informative study, the company elected not to pursue the opportunity. However, that costly research proved to be an invaluable foundation for my own independent ecosystem research.

After twelve years with that company, I left big corporation life forever, forming my own independent consulting firm. Activities as a Management Systems Consultant, Land Use Planner and Architectural Designer led to a mix of successful business ventures and guest lecturer activities on University of California campuses, The American Management Association and others. This, in turn, led, over many years, to an independent research resulting in The OMEGA Matrix.

ACKNOWLEDGEMENTS

In my checkerboard career, it was not always easy to identify with people in high places who were willing to step outside of the conventional wisdom in dealing with certain problems and techniques. Not only did I learn from the people listed, but found them stepping forward in meaningful support of my objectives at various points in time. While I do not hold them accountable for the approach to the OMEGA matrix, their collective wisdom has been a marked influence. They are as follows:

Col. William Barksdale, Commanding Officer, U.S.Air Force. J.H.Kindelberger, President, North American Aviation.

Major General Victor Bertrandias, Deputy Inspector General,USAF. Charles F. Thomas, Corporate Executive, Lockheed Aircraft. Willis Hawkins, Corporate Executive, Lockheed Missiles & Space. George Haslein, FAIA, Head, Architecture and Engineering, UCSLO.

I am also indebted to the authors of the selected bibliography for their continuing inspiration…and to the many professional associates and students who, though not directly involved in the OMEGA matrix effort, nonetheless contributed to my thinking more than they know.

Too, I wish to thank the iUniverse people for the degree to which their editorial comment helped me realign the presentation of a number of important considerations.

The author may be contacted at:

Ralph Miner, P.O.Box 221730, Carmel, CA 93922 or 831-624-5593 and rhminer7@aol.com

TEN BIBLIOGRAPHY SELECTIONS

Peter Drucker. *The New Realities*
Henri Bergson. *Time and Free Will*
Jeremy Rifkin. *Entropy*
Jeremy Rifkin. *Time Wars*
Alvin Toffler. *The Third Wave*
Bertrand Russell. *Wisdom of The West*
Robert A. Millikan. *Autobiography of Robert A. Millikan*
John Dewey. *Theory of Valuation*
Laurie Nadel. *Sixth Sense*
World Future Sociey. *Environmental Issues and Sustainable Futures*